Seismic Design and Assessment of Bridges

GEOTECHNICAL, GEOLOGICAL AND EARTHQUAKE ENGINEERING

Volume 21

For further volumes:
http://www.springer.com/series/6011

Andreas J. Kappos • M. Saiid Saiidi
M. Nuray Aydınoğlu • Tatjana Isaković
Editors

Seismic Design and Assessment of Bridges

Inelastic Methods of Analysis and Case Studies

Springer

Editors
Andreas J. Kappos
Department of Civil Engineering
Aristotle University of Thessaloniki
Thessaloniki, Greece

M. Nuray Aydınoğlu
Department of Earthquake Engineering
Kandilli Observatory and Earthquake
Research Institute
Boğaziçi University
Çengelköy – Istanbul
Turkey

M. Saiid Saiidi
Department of Civil and Environmental
Engineering
University of Nevada
Mail Stop 258
Reno, NV, USA

Tatjana Isaković
Faculty of Civil and Geodetic Engineering
University of Ljubljana
Jamova 2
Ljubljana, Slovenia

ISBN 978-94-007-3942-0 ISBN 978-94-007-3943-7 (eBook)
DOI 10.1007/978-94-007-3943-7
Springer Dordrecht Heidelberg New York London

Library of Congress Control Number: 2012936492

Printed on acid-free paper

Springer is part of Springer Science+Business Media (www.springer.com)

Preface

Bridges are the most critical component of transport systems with respect to several criteria, including those related to earthquake response. Furthermore, their number, and hence the size of the bridge stock exposed to seismic risk, keeps increasing; as an example, within the European Union, the motorway network expanded from 39,200 km in 1990 to 49,200 km in 1999, and the figure keeps increasing at a fast rate. Nevertheless, the literature devoted to these important engineering structures, and in particular their seismic performance, design, and assessment, is quite limited compared to that related to building structures. Meanwhile, assessment of seismic performance of existing bridges, some of which clearly fail to satisfy the requirements of modern codes, has been the focus of substantial research efforts worldwide, particularly during the recent years. As a result of such concerns, programmes for retrofitting seismically deficient bridges have originated in several European countries, notably in Italy, and similar efforts are underway in the US and Japan.

This book is intended as a contribution to the limited literature related to the seismic behaviour of bridges. It focuses on a topical issue, the use of inelastic analysis methods for the seismic assessment and design of bridges, on which substantial work has been carried out in recent years, but not been collected in a single volume. In particular the most advanced inelastic analysis methods that emerged during the last decade are currently found only in the specialised research-oriented literature, such as technical journals and conference proceedings. Hence the key objective of this book is twofold, first to present all important methods belonging to the aforementioned category in a uniform and sufficient for their understanding and implementation length, and to provide also a critical perspective on them by including selected case-studies wherein more than one methods are applied to a specific bridge, and by offering some critical comments on the limitations of the individual methods and on their relative efficiency. In this respect, this book should be a valuable tool for both researchers and practising engineers dealing with seismic design and assessment of bridges, by both making accessible (and, hopefully, comprehensible) the methods and the analytical tools available for their implementation, and by assisting them to select the method that best suits the individual bridge projects that each engineer/researcher faces. A 'guided tour' to the individual chapters of the book can be found in Chap. 1 – Introduction.

This book is the outcome of a genuinely collective effort. The idea for preparing a state-of-the-art document on the topical issue of inelastic analysis methods for bridges originated in 2007 during the first meeting of the (then) newly formed Task Group 11 "Seismic Design, Assessment, and Retrofit of Bridges" of the EAEE (European Association of Earthquake Engineering). Several members of this international group have contributed initially to the pertinent discussions and subsequently to the collection of material coming from their recent, or even ongoing, research. It was not long before it was realised that the need for a book on this topic was clear and the material that gradually accumulated formed a good basis for such a book. Discussions with the Secretary General of the EAEE Atilla Ansal and later with Springer's Senior Publishing Editor Petra Steenbergen were quite encouraging and have eventually led to the decision for producing this volume. As should be clear from comparing the Contributors' list with the TG11 membership (http://eaee-tg11.weebly.com/membership.html), this book has been written, almost exclusively, by members of the Group. Having said this, the material included here derives from research conducted all over the world, especially in North America and Europe, where inelastic analysis methods for bridges have been the subject of extensive scrutiny and research. Since all the available methods have been discussed (to the largest feasible extent) within TG11, the material presented in this book can also be considered as a consensus view of the research community on issues that remain, to a certain extent, still open to discussion and further improvement.

The editors of the book, all of them members of EAEE's TG11, wish to thank all those who participated to the several discussions that took place within the Group, as well as contributed material to it. In addition to the contributors to the individual chapters, who are listed under each chapter, the Editors would like to acknowledge the contributions of the following individuals:

- Dr Anastasios Kotsoglou from Democritus University of Thrace, Olympia Taskari, Doctoral Candidate at the Aristotle University of Thessaloniki (AUTh), and Karin Saxon, Executive Administrative Assistant, University of Nevada, Reno, for their assistance in preparing Chap. 2.
- Dr Göktürk Önem, from KOERI, Boğaziçi University for his assistance in preparing Chap. 3.
- Themelina Paraskeva, Doctoral Candidate at AUTh, Mauro Popeyo Nino Lazaro, from UNAM, Mexico City, and Drs Giuseppe Perrone and Salvatore Sofia from University of Basilicata, for their assistance in preparing Chap. 4.

Clearly, several other researchers have produced results that are included in the various chapters of this book; their contribution is acknowledged in the usual way, by reference to their pertinent publications in the international literature.

All chapters of the book have been subjected to external peer review; special thanks are due to the reviewers Prof. Paolo Pinto, Prof. Bozidar Stojadinovic, and Dr Martin Koller.

Last but not least, the editors would welcome comments from the readers of this book. As mentioned earlier, issues that are still open to discussion and further improvement are addressed in this volume, hence any comments made by the reviewers will be duly taken into account in a second edition that might emerge in the future.

Thessaloniki Andreas J. Kappos

Contents

Contributors

Antonio Arêde Departamento de Engenharia Civil, Faculdade de Engenharia, University of Porto, Porto, Portugal

M. Nuray Aydınoğlu Department of Earthquake Engineering, Kandilli Observatory and Earthquake Research Institute, Boğaziçi University, Çengelköy – Istanbul, Turkey

Donatello Cardone Deptartment of Structures, Geotechnics and Applied Geology, University of Basilicata, Potenza, Italy

Pedro Delgado Escola Superior de Tecnologia e Gestão, Instituto Politécnico de Viana do Castelo, Viana do Castelo, Portugal

Mauro Dolce Seismic Risk and Post-Emergency Office of the Italian Department of Civil Protection, Rome, Italy

Matej Fischinger Faculty of Civil and Geodetic Engineering, University of Ljubljana, Ljubljana, Slovenia

Tatjana Isaković Faculty of Civil and Geodetic Engineering, University of Ljubljana, Ljubljana, Slovenia

Andreas J. Kappos Department of Civil Engineering, Aristotle University of Thessaloniki, Thessaloniki, Greece

Stavroula Pantazopoulou University of Cyprus (on unpaid leave from Democritus University of Thrace), Nicosia, Cyprus

Gokhan Pekcan Department of Civil and Environmental Engineering, University of Nevada, Reno, NV, USA

Rui Pinho Department of Structural Mechanics, University of Pavia, Pavia, Italy

M. Saiid Saiidi Department of Civil and Environmental Engineering, University of Nevada, Reno, NV, USA

Anastasios Sextos Department of Civil Engineering, Aristotle University of Thessaloniki, Thessaloniki, Greece

Nelson Vila Pouca Departamento de Engenharia Civil, Faculdade de Engenharia, University of Porto, Porto, Portugal

Chapter 1
Introduction

Andreas J. Kappos

Modern codes for seismic design of buildings, bridges, and other civil engineering structures offer to the designer the choice between elastic and inelastic analysis methods, i.e.

- 'Traditional' methods wherein design is based on the results of a series of *elastic* analyses that provide linear action effects (moments, shears, axial loads) which are reduced by a global force reduction factor (q-factor in Europe, R-factor in the US) that depends on the overall ductility and overstrength capacity of the structure.
- Displacement and/or deformation based methods, wherein inelastic deformation demands in the structure are estimated for a given level of the earthquake action (typically expressed in terms of the displacement of a control point of the structure) with the aid of a series of *inelastic* (i.e. material nonlinear) analyses of either the static or dynamic type. These demands are then checked against the corresponding deformation capacities of the critical structural elements.

Over the last two decades, researchers and engineers have gradually shifted towards the *performance-based assessment and design* concept, wherein inelastic deformation demands are (preferably) directly obtained from the aforementioned nonlinear response analysis of the structure. The safety verification then involves comparing these demands against the deformation capacities (acceptance criteria) to verify the performance of the structure with respect to a given performance objective (e.g. allowable member rotation for ensuring life safety under a ground motion having an appropriately selected probability of occurrence). It is worth noting that since inelastic analysis presupposes knowledge of the strength of members (which is not required in elastic analysis, unless member stiffness is

A.J. Kappos (✉)
Department of Civil Engineering, Aristotle University of Thessaloniki,
54124 Thessaloniki, Greece
e-mail: ajkap@civil.auth.gr

A.J. Kappos et al. (eds.), *Seismic Design and Assessment of Bridges: Inelastic Methods of Analysis and Case Studies*, Geotechnical, Geological and Earthquake Engineering 21,
DOI 10.1007/978-94-007-3943-7_1,

formally determined as the secant value at yield), applying such an analysis to a reinforced concrete structure requires an initial design to be carried out, from which member reinforcement will be selected. Hence, this type of analysis is more suitable for the assessment of existing structures than for the design of new ones.

This book focuses on the use of inelastic analysis methods for the seismic assessment and design of bridges, for which the work carried out so far, albeit interesting and useful, is nevertheless clearly less than that for buildings. It should be pointed out that although some valuable literature, including books like those by Priestley et al. (1996) and Chen and Duan (2003), is currently available, the most advanced inelastic analysis methods that emerged during the last decade are currently found only in the specialised research-oriented literature, such as technical journals and conference proceedings. Hence the key *objective* of this book is two-fold, first to present all important methods belonging to the aforementioned category in a uniform way and to a sufficient for their understanding and implementation length, and to provide also a critical perspective on them by including selected case-studies wherein more than one methods are applied to a specific bridge, as well as by offering some critical comments on the limitations of the individual methods and on their relative efficiency. The last point is a crucial one since, as will be made clear later in this book, 'simplified' inelastic methods, notably those based on static (as opposed to dynamic, response-history) analysis, were recently made quite sophisticated and able to handle problems hitherto tackled solely by dynamic analysis, but (perhaps inevitably) they also became quite complex, while the computational effort involved has increased to an extent that makes questionable the benefits (if any) gained by using these static methods in lieu of the dynamic one. Hence the material included in the book constitutes an aid for seismic design and assessment of bridges, presenting both the methods (Chap. 3) and the analytical tools (Chap. 2) available for their implementation, and providing guidance (Chap. 5) for selecting the method that best suits the specific bridge project at hand.

The main part of this book consists of Chaps. 2, 3 and 4. In ***Chap. 2*** the analytical tools necessary for the implementation of inelastic methods for bridges are presented. The chapter starts with available models for the bridge deck and their role in seismic assessment, addressing not only elastic modelling of the deck (which is the most commonly adopted approach) but also far less explored issues like the verification of deck deformation demands in cases that inelastic behaviour of the deck is unavoidable. Then the topic of modelling bearings and shear keys is presented, which is of paramount importance in the case of bridges, logically followed by the related issue of seismic isolation and energy dissipation devices; modelling of all commonly used isolation and dissipation devices is discussed and practical guidance is provided. The longest section in this chapter is devoted to inelastic modelling of different types of bridge piers, which is not surprising if one notes that piers (especially single-column or multi-column ones) are the bridge components wherein seismic energy dissipation takes place in non-isolated structures. All types of inelastic models for members, with emphasis on reinforced concrete columns, are presented in a rather detailed way, including both lumped plasticity and distributed plasticity models (distributed flexibility elements and fibre models of different types). Several examples of application of the previously mentioned models to bridges of varying complexity are provided and critically discussed. The last two sections of the chapter (Sects. 2.6 and 2.7) deal with modelling of the foundation of bridges and its interaction with the ground. Simple and more sophisticated models for abutments and (surface and deep) foundation members are provided, followed by models for the surrounding ground, with emphasis on the embankments that often play a crucial role in the seismic response of bridges, in particular short ones. Soil-structure interaction modelling of bridges is presented in both its commonly used forms, i.e. linear, as well as nonlinear soil-foundation-bridge interaction analysis in the time domain. These last sections of the chapter also include a brief overview of a major topic, i.e. the characteristics of seismic ground motion which is used as input for the analysis, the detailed presentation of which is beyond the scope of the book.

Chapter 3 is the core of the book, in the sense that it presents in a uniform way the available inelastic analysis methods for the seismic assessment and design of bridges. Since inelastic response-history analysis has long been used for bridges without substantial changes or developments during the last decade, it is presented in a rather brief and concise manner, leaving aside details of the numerical integration of the equations of motion that can be easily found in structural dynamics textbooks (e.g. Chopra 2006). On the contrary, inelastic static (pushover) methods, which have been the focus of extensive research in the recent years, particularly in the direction of extending them to structures with significant higher mode effects (a typical example being the transverse direction of many bridge types), occupy the largest part of the chapter. It should be noted that the methods described do not encompass all variations of pushover analysis techniques that deal with approximate ways for treating higher mode effects; instead, only those methods that have been specifically applied (after proper tailoring) to bridges have been selected. Having said this, the authors believe that the methods presented in Chap. 3 practically include all different approaches to the problem, and very few techniques are left outside on the grounds that they were used solely for buildings. To allow for a more rational presentation, the different methods are presented not in a strict chronological order, but by classifying them into 'single-mode' and 'multi-mode' pushover analysis procedures. In the first category the now well-known N2 method is described in sufficient length as a typical representative. Within the 'multi-mode' pushover category a distinction is made between

- procedures involving a series of *individual-independent* modal pushover analyses (among which the modal pushover analysis technique adapted to the needs of bridges is presented in detail)
- multi-mode procedures based on adaptive incremental implementation of response spectrum analysis for simultaneous modal pushover analyses (the IRSA method is described in detail), and
- multi-mode procedures based on single-run pushover analysis with modal-combined adaptive seismic load or displacement patterns (the 'Adaptive Capacity Spectrum Method' is presented as the most recent version of this approach)

Finally, a brief overview of an approach presented some time ago but not elaborated further ever since, the 'Modal Adaptive Nonlinear Static Procedure (MANSP)' is provided. For all major approaches to pushover analysis, which, as noted earlier, are only available in the specialised literature, an effort is made to describe them in such detail that would permit their implementation by the reader, provided, of course, that he/she has the appropriate analytical tools (cf. Chap. 2) and the expertise to use them. Practical application is also facilitated considerably by the information provided in the next chapter.

In ***Chap. 4***, the methods presented in the previous chapter are applied to specific case-studies, involving bridges with different length and configuration. The chapter starts with a critical discussion of the basic parameters that influence the applicability of pushover methods. Then, a number of case-studies are presented in a rather uniform and detailed way; they were selected among those available with a view to including at least one application of each category of methods described in the previous chapter and (wherever feasible) to applying two or more 'simplified' methods to the same bridge structure. In addition to a number of pushover analyses, all case-studies include also response-history analysis of the inelastic response of the bridge, which serves as a reference for evaluating the results of the various approximate (static) procedures. It is worth pointing out that in the case studies, in addition to the four pushover methods described in detail in Chap. 3, some other variants of the key approaches are also used and comparatively evaluated, so that at the end a more global picture of practically all analysis and assessment techniques available for bridges is provided. To allow for an even broader view on the issues involved and put the purely analytical methods into the proper perspective, the final section of Chap. 4 presents an experimental evaluation of analytical methods, i.e. results from analytical methods (response-history, as well as pushover) are compared with those from the shaking table testing (using three shaking tables) of a 1:4 scale bridge model.

The key conclusions regarding the range of applicability and the relative performance of the analytical assessment methods presented in the book are summarised in ***Chap. 5***, and some specific recommendations are made regarding the selection of analytical model and analysis method in a practical context. It must be emphasised that it is *not* the aim of this book to suggest the 'best' method of seismic assessment of bridges, since none of the existing methods would qualify as such; selection of analysis method is heavily problem-dependent and to a large extent available software-driven, while, in practical applications, code requirements also affect the selection of analytical tools.

References

Chen WF, Duan L (eds) (2003) *Bridge Engineering: Seismic Design (Principles and Applications in Engineering)*, CRC Press, Boca Raton

Chopra AK (2006) The dynamics of structures: theory and applications to earthquake engineering, 3rd edn. Prentice-Hall, Upper Saddle River

Priestley MJN, Seible F, Calvi GM (1996) Seismic design and retrofit of bridges. John Wiley and Sons, Inc (Editor), New York

Chapter 2
Modelling of Bridges for Inelastic Analysis

M. Saiid Saiidi, Antonio Arêde, Donatello Cardone, Pedro Delgado, Mauro Dolce, Matej Fischinger, Tatjana Isaković, Stavroula Pantazopoulou, Gokhan Pekcan, Rui Pinho, and Anastasios Sextos

2.1 Introduction

Engineers rely on mathematical models to help analyze and design structures while being aware that the actual structure might behave differently than that assumed in the mathematical model. The ability of engineers to reduce the gap between the actual and assumed behaviour of structures has increased substantially due to new knowledge about material behaviour and the rapid development of computing tools. Earthquake engineering of bridges is particularly challenging because of the variety of components and parameters that affect the bridge seismic response. Inelastic analysis of bridge response has the potential of being a powerful approach towards realistic estimate of the actual behaviour of bridges. However, the power and

M. Saiid Saiidi (✉)
Department of Civil and Environmental Engineering, University of Nevada, Reno, NV 89557, USA
e-mail: saiidi@unr.edu

A. Arêde
Departamento de Engenharia Civil, Faculdade de Engenharia, Universidade do Porto, Rua Dr. Roberto Frias, s/n, 4200-465 Porto, Portugal
e-mail: aarede@fe.up.pt

D. Cardone
Department of Structures, Geotechnics and Applied Geology, University of Basilicata, Viale dell'Ateneo Lucano 10, 85100 Potenza, Italy
e-mail: donatello.cardone@unibas.it

P. Delgado
Escola Superior de Tecnologia e Gestão, Instituto Politécnico de Viana do Castelo, Avenida do Atlântico, 4900-348 Viana do Castelo, Portugal
e-mail: pdelgado@estg.ipvc.pt

M. Dolce
Department of Civil Protection, Via Ulpiano 11, 00193 Rome, Italy
e-mail: mauro.dolce@protezionecivile.it

A.J. Kappos et al. (eds.), *Seismic Design and Assessment of Bridges: Inelastic Methods of Analysis and Case Studies*, Geotechnical, Geological and Earthquake Engineering 21, DOI 10.1007/978-94-007-3943-7_2, versatility of inelastic analysis also makes the results particularly sensitive to improper application. The material presented in this chapter is intended to inform researchers and designers about available modelling methods and important parameters to help ensure that inelastic analysis of bridge earthquake response is conducted properly.

2.2 Superstructure (Deck)

This section addresses modelling issues of bridge decks occurring in dynamic analysis of bridges. Generally, today's finite element programs contain libraries of elements that can support a range of options regarding the accuracy of representation of superstructure's geometry and loading – however, practitioners rarely resort to detailed 3D solid elements for that purpose, in order to minimize the modelling effort and computational time. Rather, the degree of detail used in idealization (modelling) of the superstructure in practical bridge analysis is often associated with the type of loading applied.

In gravity load analysis the geometric complexity of the deck is usually represented in the computer model to greater detail, as compared to that used for transverse loads such as in analyzing ground-motion induced, seismic response. This is probably justified on grounds of the perceived discrepancy in the amount of flexural stiffness of the deck in transverse and longitudinal deformation as compared to that of the more flexible piers, bearings, restrainer hardware and embankments, which undertake accordingly the large portion of the bridge's

M. Fischinger • T. Isaković
Faculty of Civil and Geodetic Engineering, University of Ljubljana,
Jamova 2 1000 Ljubljana, Slovenia
e-mail: matej.fischinger@ikpir.fgg.uni-lj.si; tisak@ikpir.fgg.uni-lj.si

S. Pantazopoulou
Department of Civil and Environmental Engineering, University of Cyprus,
Aglantzia, Cyprus
e-mail: pantaz@ucy.ac.cy; pantaz@civil.duth.gr

G. Pekcan
Department of Civil and Environmental Engineering, University of Nevada,
Reno, 1664 N. Virginia Street, Reno, NV 89557, USA
e-mail: pekcan@unr.edu

R. Pinho
Department of Structural Mechanics, University of Pavia, Via Ferrata 1, 27100 Pavia, Italy
e-mail: rui.pinho@unipv.it

A. Sextos
Department of Civil Engineering, Aristotle University of Thessaloniki,
54124 Thessaloniki, Greece
e-mail: asextos@civil.auth.gr

displacement demand. For this reason, detailed modelling of the deck stiffness is only considered of use in evaluating the two-way action of the deck, in assessing the effects of vertical service loads and self weight.

In dynamic analysis for evaluation of seismic response, be it linear-elastic or nonlinear, a common approach to deck modelling is to represent the superstructure by a spine of linear beam-column elements with lumped nodal masses, spanning between successive nodes along the deck's length. The spine follows the centre of gravity of the cross section along the length of the bridge. The deck geometry is used to evaluate equivalent sectional frame-element properties, often using standard flexural analysis in order to evaluate the gross, or cracked sectional stiffness, *EI*, of the deck acting as a single beam. Note that superstructure bending can be expected to cause or enhance already existing cracking from gravity loads and live loads in reinforced concrete structures. In such cases the usual practice is to assume effective or cracked stiffness properties for the moment of inertia of the entire cross section about the transverse or *y* axis. While detailed cracked-section stiffness analyses can be performed for each girder or the complete superstructure, it is often sufficient to calculate the gross section stiffness I_g and reduce it to $I_{eff} = 0.5 \cdot I_g$ for reinforced concrete, while assuming no stiffness reduction in prestressed decks. The torsional rigidity *J* for spine elements can also be determined from standard mechanics principles and can be considered as fully effective as long as the torsional cracking moment is not exceeded, at which point the torsional stiffness significantly reduces. The modelling procedure detailed above corresponds to the assumption of a plane-sections response, whereby longitudinal and transverse behaviour only depend upon the position along the longitudinal axis – a uniform distribution of all effects occurs in the transverse direction in this type of idealization, and this is why it is inappropriate for gravity load analysis.

California Department of Transportation (CALTRANS) endorses the approach of modelling the bridge deck using a spine of 3D frame elements (6 degrees of freedom at each node, and a minimum of three nodes in each span – at quarter points), with pertinent assignment of lumped mass. Recommended stiffness properties of the superstructure are based on the assumption of un-cracked cross sections since the superstructure is expected to respond linearly to seismic loadings. Alternatively, an equivalent orthotropic elastic slab system of uniform depth is proposed, having the same overall stiffness in each primary direction with that of the longitudinal and transverse beams of the original deck. In seismically isolated bridges, the superstructure is modelled as a rigid body and is designed for forces calculated using the effective stiffness of the isolation system at the design displacement.

An alternative to the single orthotropic slab model for the deck is a hybrid shell – beam finite element assembly. In this approach, the deck slab is modelled with rectangular shell elements. Girders are modelled with 3D frame elements connected to the shell elements at each node with vertical restraints in order to account for the contribution of the girders' weak-axis moment of inertia to the superstructure stiffness for transverse loading. In the longitudinal direction, the end of the girder is attached to a spring representing the elastomer's lateral stiffness.

Consideration of the vertical component of ground motion and the coupling between longitudinal, vertical and transverse response that occurs in curved bridges under any kind of excitation, underscores the limitation of the single-spine 3D frame model when applied to general purpose dynamic analysis of bridge superstructures. More detailed than single-element beam-type idealization is finite element discretization or other, simplifying modifications to this approach such as finite strip, folded plate, and finite segment analysis. For example, a box-girder bridge superstructure may be represented by a mesh of two-dimensional shell or plate elements. A convenient compromise is the so-called grillage-type model, which idealizes the structure as a space-frame configuration comprising a network of longitudinal and transverse beams. It represents a balance between accuracy and practicability of modelling approach, such as minimum effort for input preparation, minimal computational effort and meaningful interpretation of output. Details for grillage-type modelling of various deck types are outlined in the following section.

2.2.1 Deck Types, Sectional Layouts and Properties

The simplest deck type is that of a *compact slab*, which, from among the various section types, possesses the lowest flexural stiffness. In a beam-type approximation for either longitudinal or transverse action, interaction between the two directions is a significant aspect that need be accounted for in any model short of 3D solid-element representation. Therefore, in the grillage approximation the deck is idealized as an assembly of basic modelling units, each comprising a rectangular horizontal frame with diagonal elements; the latter are meant to account for the interaction between longitudinal and transverse deck action owing to Poisson's effects, and cannot be neglected in the absence of primary girders. Properties of the associated elements are listed in Table 2.1, where, I_x and I_y are the flexural moments of inertia along the longitudinal and transverse directions of the bridge, respectively, J_x and J_y are the torsional moments of inertia along the longitudinal and transverse directions, I_d the diagonal element flexural stiffness, ν is the Poisson's ratio, L_x and L_y the breadth and length of the grillage unit, and t is the deck thickness.

2.2.1.1 Beam and Slab Bridge Decks

In the idealization of this deck system the longitudinal axes of the analytical model are located on the axes of the girders, whereas the slab is idealized with a series of transverse strips, each being modelled by a beam. Member properties for the analytical model are listed in Table 2.1, where, I_x is the flexural moment of inertia with respect to the centroidal axis of the combined section, L_x is the width of the idealized transverse "beam", L_y the width of the idealized longitudinal beam (equal to the girders' spacing). The expressions above neglect the Poisson's effect and assume that in the transverse direction flexural rigidity is approximately equal to the torsional rigidity (valid for square configuration of grillages).

Table 2.1 Properties of equivalent members of decks for deck modelling[a]

Deck section	Slab	Girders	Cellular	Voided
Diagonals	$J_d = \left[\frac{v(L_x^2 + L_y^2)^{1.5}}{L_x L_y}\right] \cdot \frac{t^3}{24(1-v^2)}$			
Shear area			$A_y = \frac{12 L_x E_s}{P_y G_s} \cdot \cdot [n_s I_3 (12 I_1 I_2 + P_x I_1 I_3 + n_s P_x I_2 I_3)] / \left[4 P_y I_1 I_2 + 4 n_s P_y I_2 I_3 + 12 H^2 I_1 I_2 + n_s P_y^2 I_3^2\right]$	$A_y = F_1 \cdot t \cdot L_x$
J_y	$J_y = \left[\frac{E_s L_x (1-3v)}{G_s}\right] \cdot \frac{t^3}{24(1-v^2)}$	$J_y = L_x \frac{E_s}{G_s} \frac{t^3}{12}$	$J_y = 2L_x \frac{A_2^2}{b \oint \frac{ds}{n_s t}}$	$J_y = \left(1 - 0.84\left(\frac{t_v}{t}\right)\right) L_x \frac{t^3}{6}$
J_x	$J_x = \left[\frac{E_s L_y (1-3v)}{G_s}\right] \cdot \frac{t^3}{24(1-v^2)}$	$J_y = J_{g,x} + L_y \frac{t^3}{6}$	$J_x = L_y \frac{A_1^2}{b \oint \frac{ds}{n_s t}}$	$J_x = \left(1 - 0.84\left(\frac{t_v}{t}\right)\right) L_y \frac{t^3}{6}$
I_y	$I_y = \left[L_x - v\frac{L_y^2}{L_x}\right] \cdot \frac{t^3}{24(1-v^2)}$	$I_y = L_x \frac{t^3}{12}$	$I_y = L_x \frac{t_1^3}{12} + n_s L_x \frac{t_2^3}{12} + L_x (t_1 h_1^2 + n_s t_2 h_2^2)$	$I_y = \frac{L_x t^3}{12} \left[1 - 0.95\left(\frac{t_v}{t}\right)\right]^4$
I_x	$I_x = \left[L_x - v\frac{L_x^2}{L_y}\right] \cdot \frac{t^3}{24(1-v^2)}$	$I_x = I_{g,x} + I_{s,x}$	$I_x = I_{g,x} + I_{s,x}$	$I_x = \frac{P_y t^3}{12} - \left(n_s \frac{t_v^4}{64}\right)$
Section notation	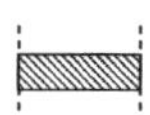	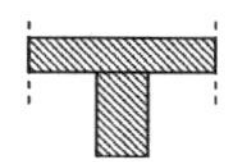	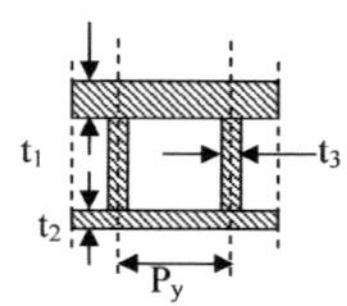	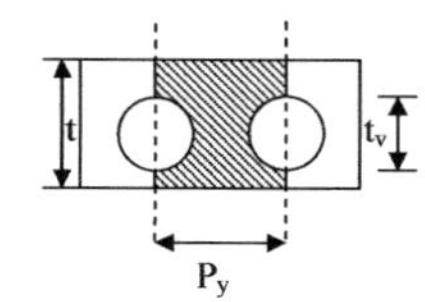

[a]Increased transverse stiffness can be obtained at locations where transverse diaphragms are present in the superstructure. Transverse diaphragms can be modelled with transverse grillage beam elements with characteristics derived from T or I sections and an effective flange width of eight times the slab thickness on either side of the diaphragm

2.2.1.2 Cellular or Box-Section Bridge Decks

Idealization procedures in the longitudinal direction are similar to those described in the preceding case for slab-and-beam bridge decks; idealized longitudinal members are taken as I-beams placed along the axes of beam-webs of the structure (centrelines between adjacent cells). The top and bottom slabs of the deck are considered as top and bottom flanges of the idealized "beam". Since the centre of gravity of the deck is generally in a higher position with respect to the level of the bearings, a rigid vertical link should be introduced connecting the deck axis and the bearings. Torsional rigidities are based on the whole deck section rather than on the individual idealized beams. Intermediate webs and overhangs are excluded in this calculation due to their negligible contribution; thus, J_x and J_y are given in terms of the area enclosed by the median line of the closed cross section, A_1 (Table 2.1), *2b* is the width of the entire deck, n_s the shear moduli ratio of bottom and top deck flanges, and *t* the section's thickness. The torsional inertia thus estimated is then assumed uniformly distributed across the deck in proportion to the width of the idealized members. Flexural inertia and shear area for idealization in the transverse direction are listed in Table 2.1, where E_s is the modulus of elasticity of the deck slab, G_s is the shear modulus of the deck slab, I_1 and I_2 are the moments of inertia per unit length of the top and bottom flanges, I_3 is the moment of inertia per unit length of the real web.

2.2.1.3 Voided-Slab Decks

Idealization is based on the same principles as used in cellular bridges. Note that voids reduce significantly the transverse rigidity of the superstructure but the shear-lag effect which is prominent in Cellular-type decks is not as significant in this case. The necessary expressions for idealizing the deck as a system of longitudinal and transverse beams are also given in the table.

2.2.1.4 Effect of Plan Curvature

In the presence of deck curvature, response to longitudinal or vertical action becomes coupled to response in the transverse direction. The practical significance of this coupling depends on whether the deck functions as an *open* or a *closed* system. Open systems are curved deck types that lack transverse bracing (very small torsional stiffness), thereby being susceptible to torsional loads. A general practice is to assume uncracked torsional stiffness properties if the anticipated torques due to seismic loads are low (below the cracking limit), but only 5% of that value if the cracking torque limit is exceeded. This is particularly important in highly curved superstructures, because a result of the decks' geometric curvature is the natural development of a distributed torque associated with the diagram of longitudinal flexural moments.

Thus, this aspect of the behaviour need be properly reflected in the analytical model. Closed systems can be analyzed as single-section curved beams, or using the multi-beam idealization discussed previously for straight decks.

2.2.1.5 Effect of Prestressing on Component Verification Issues

The general practice is to use gross section properties when modelling the flexural rigidity of prestressed deck sections, if the estimated seismic demands for the deck are below cracking, whereas the average cracked stiffness is used in cases where the anticipated demands exceed the cracking limit. Note however, that the prestressing force affects primarily the deck's stiffness through its influence on the extent of flexural cracking. Therefore, under load reversals the stiffness after cracking greatly depends on the location of the prestressing cable's centroid and the direction of bending, an aspect that may only be represented through non-linear modelling of the deck member behaviour even in the pre-yielding range of response; linear elastic idealization with average values cannot reproduce the implications of stiffness variation imparted on the prestressed section by moment reversal. Therefore, no uniform stiffness reduction is recommended for prestressed concrete box girder sections to account for the effect of cracking, but rather a detailed moment-curvature analysis may be needed in order to assess the actual deck section's flexural stiffness for nonlinear response history analysis (NRHA).

2.2.2 Role of Deck Modelling in Seismic Assessment

Either conducted through standard pushover analysis (SPA) or through NRHA in assembling the bridge model the superstructure is usually considered to remain in the elastic range. Due to its in-plane rigidity, it is often assumed to move as a rigid body under seismic loads, and the entire modelling problem is reduced to modelling of the bents with geometric constraints simulating the rigid superstructure. Nonlinear modelling is then restricted to represent joints between superstructure segments, connections with supporting bents, and to the assessment analyses of older bridges, where the superstructure may not be protected against inelastic action by capacity design principles. In general, the standard pushover analysis can provide satisfactory predictions to the maximum displacement responses of straight bridges in the longitudinal direction (residual displacements tend to differ even in this case, due to the different hysteretic models used in detailed nonlinear response history analysis as compared to the equivalent single-degree-of-freedom (SDOF) approximation of the SPA). In some cases, response of long bridges in the longitudinal direction is complicated by the impact between adjacent spans. When subjected to seismic loading, movement or expansion joints allow separate frames to develop their own characteristic dynamic response and to modify this individual dynamic response through complex interactions that occur in the restraining devices installed in the movement joints. For an elastic model or a plane frame inelastic static model to be applicable, adjacent frames should have comparable stiffness and mass characteristics in order for the structure to respond in its fundamental mode of vibration. Unbalanced frame geometry will lead to larger relative displacements and transfer of increased seismic force from one frame to the other. Thus, if the linear elastic displacements to the design earthquake exceed the expansion joint width, the impact between successive spans should be modelled as well, in the framework of NRHA.

But the difference between SPA and NRHA is particularly relevant when the bridge response is analyzed in the transverse direction. Note that in this case superstructure stiffness greatly affects the seismic behaviour of the bridge: excessive deck stiffness will equalize displacement demands on the piers leading to damage concentration in shorter piers and end abutments and embankments; flexible decks will lead to decoupled pier behaviour. Thus, it is important to faithfully represent in the model the bridge deck flexibility, particularly in the case of long structures, for two practical reasons: (a) in order to capture higher mode contributions which can become significant for irregular systems, (b) so that the distribution of seismic forces, which are applied along the length of the deck, follow the pattern of lateral deflection occurring in the deck due to its own compliance. A first step in order to obtain a reliable pattern of lateral load distribution, which may then be used in conducting SPA on long or irregular decks is to perform first a standard deflection analysis of the deck, whereby the deck model is loaded in the transverse direction by the superstructure's own gravity loads while accurately representing the support restraints against transverse translation or rotation, if any exist. In this analysis standard linear elastic modelling with gross sectional properties as listed in Table 2.1 are sufficient. Where vertical superstructure flexibility reduces the fixity of the top boundary conditions of columns and piers, a reduction in the stiffness coefficient can be determined. Stiffness contribution of non-structural elements is generally neglected; however, their contribution to inertia forces (masses) should be considered.

2.2.3 *Effects of Skew and Curvature in Plan*

Other issues pertaining to deck's seismic demand assessment are, (a) the effects of highly skew supports on bridge response in both transverse and longitudinal direction, (b) contribution of rotational inertia of the superstructure mass in transverse vibration of the bridge (particularly relevant in the case of tall piers and massive deck cross sections) and (c) contribution of ramp structures and interactions between frames, (d) in-plan and in-elevation curvature, (e) Ductility and plastic hinge formation in secondary deck elements (continuity slabs). Clearly these effects may only be accurately represented through 3Dimensional models that explicitly account for all the associated kinetic and kinematic interactions. Note that particularly skew bridges have a tendency to rotate in the horizontal plane even under non-seismic loading. Skewness contributes to high shear forces in central pier columns and to an increase in deck displacement. In general, longitudinal shaking produces transverse components of force, and vice versa. Transverse shaking causes one end of a deck panel to bear against the adjacent deck while the opposite end swings free in response to the seismic loading, resulting in a ratcheting type of behaviour under cyclic loading. The non-uniformity of movement joints behaviour (opening-pounding) necessitates their modelling using their exact geometry. To effectively reflect in the analytical model the stiffness changes of restrainers and shear keys that occur during cyclic loading, nonlinear spring elements with pertinent skeleton and hysteresis rules are usually required. Nonlinear impact at the points of contact of successive skew decks must be reflected in either longitudinal or transverse response analysis through pertinent modelling of the interaction. To enable nonlinear contact behaviour, NRHA appears to be the preferred option. For this type of problem simplifications using either rigid deck or single-spine beam-type representation of the deck are inappropriate as they generally fail to reproduce the axial forces induced in the piers by the earthquake.

Rather, a two-dimensional grillage model for the superstructure can be effective as long as a sufficient number of transverse restraints through idealized "beams" spanning between longitudinal girders are provided in the model, and the applied loads are distributed to all nodes over the bridge deck area. As described in the relevant section, torsional stiffness properties of the deck superstructure are distributed equally to the number of longitudinal grillage elements.

As far as in-plan curvature is concerned, the curvature can increase the transverse stiffness of the system depending on its continuity. The forces in the expansion joints are high and the movements potentially large, and hence the joints are very prone to damage caused by multiple impacting in both torsional and translational modes of response. A curved bridge without expansion joints would be extremely stiff in the horizontal plane, particularly if fully restrained at the abutments.

Secondary deck elements such as continuity slabs form the extension of the deck and are part of the external restraining system of the bridge. They act as a tension tie when the deck contracts, or as a compression strut when the deck expands. During an earthquake, the continuity slab can be either loaded by compression or tension, due to the longitudinal movements of the deck, or even by in-plane bending, due to the transverse response of the deck. Other secondary elements such as the link slabs connect adjacent simple-span girders meant by design to reduce or eliminate cracking in the concrete deck can be modelled using the cracked stiffness properties (secant to yield).

2.2.4 Verification of Deck Deformation Demands

Uncracked element section properties are typically used in practice for verification checks. This is conservative for force-controlled members, but can lead to unsafe assessment of damage which is characterized by displacement – this is one reason why reduced (cracked) stiffness properties will moderate the error in bridge assessment. Furthermore, for the purpose of seismic demand estimation, structural period should be estimated after the application of service loads.

Recent standards and guidelines for seismic assessment define the seismic input in terms of displacement demands, where the estimation of inelastic local displacements and deformation demands throughout the structure are conducted via linear static or modal response spectrum analyses. A fundamental requirement for this estimation is the use of a realistic estimation of the global elastic stiffness which, for concrete structures, corresponds to using realistic values of the effective cracked stiffness of concrete members at yielding.

When seismic response is evaluated by comparing member deformation demands to (realistic) deformation capacities, then the demands should also be realistically estimated: if member stiffness is taken equal to the default value of $0.5E_cI_c$, currently recommended for force- and strength-based seismic design of new buildings by Eurocode 8 (EC8) and US codes, then member seismic deformation demands may be seriously underestimated.

The best way to realistically estimate the effective elastic stiffness of the shear span of a concrete member (moment-to-shear ratio at the end, $L_s = M/V$) in a bilinear force-deformation model under monotonic loading, is to use the value of the secant stiffness of the shear span at member yielding $\mathrm{EI_{eff}} = \mathrm{M_yL_s/3\theta_y}$ where M_y is the value of the yield moment in the bilinear M-θ model of the shear span and θ_y that of the chord rotation at the yielding end.

Generally it is much more appropriate to conduct verification checks using detailed M-ϕ analysis of the deck sections in the critical regions, obtained after consideration of all the necessary interactions (effective flange width, prestressing, confinement, prestress losses, reinforcement slip, axial load, moment, and shear interaction) which may be estimated using web-based available freeware (e.g., RESPONSE, ANSR, ETOOLS). Recent research has shown that pier rotation capacity for piers designed according to current AASHTO requirements exceeds the value of 4% even for cases with a high axial load value (Inel and Aschheim 2004). Closed form expressions for pier rotation capacity at drift and ultimate have been developed by Biskinis and Fardis (2007):

For circular section columns of diameter D, the chord rotation at yield is given by the expression:

$$\theta_y = \phi_y \frac{L_s + a_v z}{3} + 0.0022 \max\left(0, \left(1 - \frac{L_s}{6D}\right)\right) + a_{sl} \frac{\phi_y d_b f_y}{8\sqrt{f_c}} \tag{2.1}$$

and the ultimate chord rotation is given as follows:

$$\theta_u = \theta_y + (\phi_u - \phi_y) L_{pl} \left(1 - \frac{0.5 L_{pl}}{L_s}\right) \tag{2.2}$$

where a_v is a coefficient depending on the relationship between the shear strength and the shear corresponding to flexural yielding in the member ($a_v = 1$ if the former is higher than the latter), z is the internal lever arm of section forces, a_{sl} is a coefficient depending on the adequacy of lap splices ($a_{sl} = 0$ if bar pull-out is not physically possible, otherwise $a_{sl} = 1$), ϕy is the yield curvature, d_b the bar diameter, L_{pl} is the plastic hinge length, and the other parameters have been defined previously.

For rectangular hollow section columns or T-shaped and double-T, these expressions are respectively:

$$\theta_y = 1.065\phi_y \frac{L_s + a_v z}{3} + 0.0012 + a_{sl} \frac{\phi_y d_b f_y}{8\sqrt{f_c}}, \tag{2.3}$$

and

$$\theta_u = \theta_y + 1.05(1 + 0.6a_{sl})\left(1 - 0.05\max\left(1.5, \min\left(10, \frac{h}{b_w}\right)\right)\right)(0.2)^{v} \times \left(\frac{\max(0.01, \omega')L_s}{\max(0.01, \omega)h}\right)^{1/3} f_c^{0.2} 25^{\left(\alpha\rho_s \frac{f_{yw}}{f_c}\right)} \tag{2.4}$$

where h is the section depth, and ω and ω' are the mechanical reinforcement ratios (tension /compression), ρ_s is the volumetric ratio of transverse reinforcement, and α is a confinement effectiveness factor (α=0 for unconfined sections).

If the plastic mechanism of the bridge response is known, member deformation demands (usually and conveniently estimated in the form of chord rotation demands at member ends) are related to the global displacement of the superstructure through pertinent geometric relations that are prescribed by the fundamental shape of bridge vibration (if an equivalent SDOF analysis is conducted), or otherwise estimated from detailed Finite Element Analysis. Thus, local deformation demands may be evaluated, corresponding to the bridge superstructure design displacement.

For composite concrete-steel members, the section properties should be adjusted to an equivalent concrete or steel section considering the material modular ratio. Where different concrete strengths are used in the same element (such as different concrete strengths in prestressed-concrete girders and cast-in-place concrete deck), the section properties should be transformed using a similar procedure. The composite densities should also be transformed to an equivalent concrete or steel density so that mass distribution may be dependably quantified.

2.3 Bearings and Shear Keys

The overall seismic response of bridges may be influenced significantly by the details at the abutments which may include bearings, shear keys, gap, abutment back wall, and piles. Significant soil-pile-abutment-structure interactions may take place that would lead to complex seismic response behaviour. Hence, a realistic abutment model should represent all major resistance mechanisms and components, including an accurate estimation of their mass, stiffness, and nonlinear hysteretic behaviour. In addition, bearings may be located at the piers between cap beam and the superstructure. In this section, the importance of accurate modelling of bearings and shear keys as well as simple modelling approaches are presented briefly, while other details are addressed elsewhere in this report. It is noted that representation of realistic nonlinear force-deformation relationships of bearings and shear keys is critically important for the nonlinear seismic response characterization of bridges. Various types of abutment models have been developed and used by researchers, modelling techniques and details are presented by Shamsabadi (2006), Aviram et al. (2008) and Abdel-Mohti and Pekcan (2008), which in essence can be depicted as shown in Fig. 2.6. However, the modelling requirements and techniques for bearings and shear keys depend on the type of bearing and shear key, and type of analysis. Typically, for nonlinear static and nonlinear dynamic analyses, lateral force-deformation response must be modelled explicitly in terms of a backbone (monotonic) and complete hysteretic rules, respectively.

2.3.1 Importance of Modelling of Bearings

Bridge bearings are mechanical systems which permit movements and transmit various types of loads from the bridge superstructure to the substructure. While the vertical and horizontal loads are routinely applied to bridge bearings due to gravity, traffic, wind, and other internal sources (e.g. temperature, creep, and shrinkage), seismic loads are relatively infrequent. Hence they were not considered consistently in the design of bridges until late 1980s. In spite of the role of the bearings in transmitting these loads, their importance has been often neglected (Leonhardt 1981) and the importance of accurate modelling of bearings for the seismic performance assessment of highway bridges has been overlooked. Fortunately, the level of awareness has increased since the invention and wide acceptance of "modern-type" bearings primarily for seismic applications such as PTFE sliding, elastomeric, FPS, etc. Subsequently, integrated analysis and design software that incorporate macro-level approaches to modelling of bearings have become widely available.

2.3.2 Mechanical Bearings (Steel Bearings)

Selection of proper bearing type for a bridge design is based on many considerations which include the regional seismic hazard levels. It is noted that most of existing bridges in moderate earthquake regions are more than 50 years old. The majority of these structures were designed without any consideration of earthquake forces. Hence, the so-called steel bearings (Fig. 2.1) are used in bridges that are still in service in Eastern and Central U.S., Japan, Taiwan, among others. Although current practice requires retrofitting of existing bridges using more advanced bearing types, e.g. isolation bearings, in zones of low to medium seismicity, existing steel bearings may possess sufficient strength and/or displacement capacity. Regardless, realistic force-deformation response of structural components is of fundamental importance to the process of analysis and design. With this motivation, several experimental and analytical studies have been conducted. Mander et al. introduced reliable hysteretic models for sliding, fixed, and rocker, bearings. The hysteretic

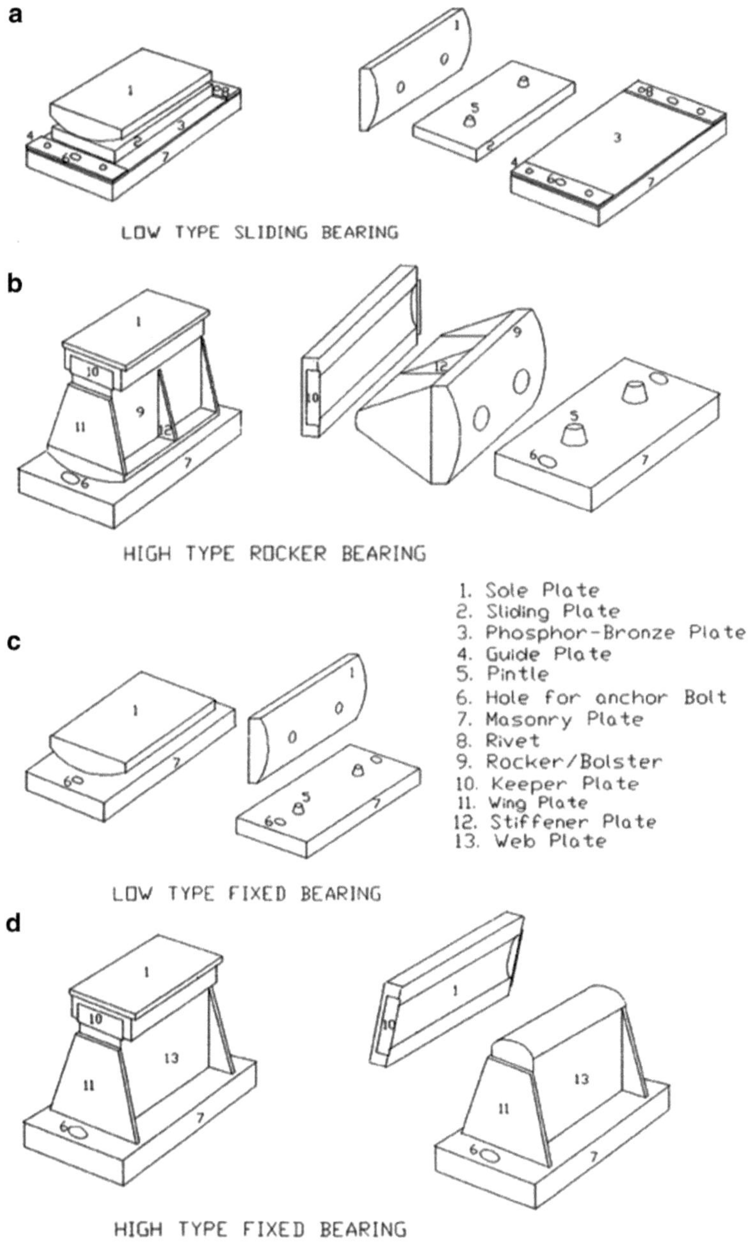

Fig. 2.1 Steel bearing types

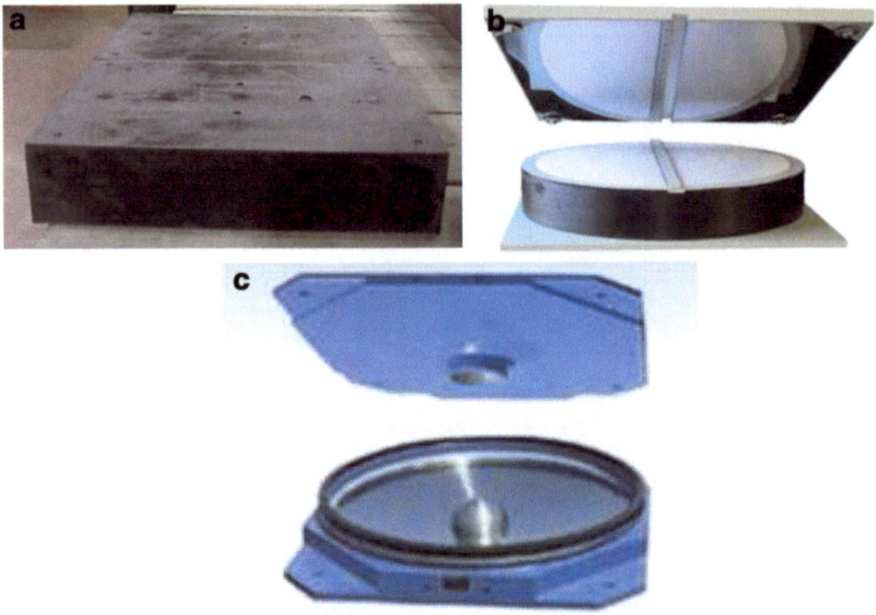

Fig. 2.2 Views of different isolators: (**a**) elastomeric, (**b**) sliding, and (**c**) friction pendulum

response of these bearing types was modelled combining a bilinear yielding element and a link element using Drain-2DX software.

2.3.3 *Modern Bearing Types*

Modern bearings are ductile, multi-rotational, and multi-directional. Accordingly, they can be identified under two broad categories, namely, elastomeric bearings and high-load multi-rotational bearings (e.g. Fig. 2.2). Plain elastomeric bearing pads (PEP), steel reinforced elastomeric bearings, and lead-rubber bearings (LRB) are the most commonly used elastomeric bearing types. Plain elastomeric bearing pads rely upon friction at the contact surface to resist building strains. The traditional non-seismic elastomeric pads used in bridges for thermal movements can provide some limited seismic protection; however their integrity during large displacements might be substantially deteriorated or even destroyed due to shearing of the elastomer or rolling of the entire bearing. Accordingly, elastomeric pads with improved seismic performance have been developed. PEPs can be modelled with an elasto-plastic hysteresis model (e.g. plastic –Wen– link element in SAP2000) with an initial elastic stiffness, k_0.

$$k_o = \frac{G \times A}{h} \tag{2.5}$$

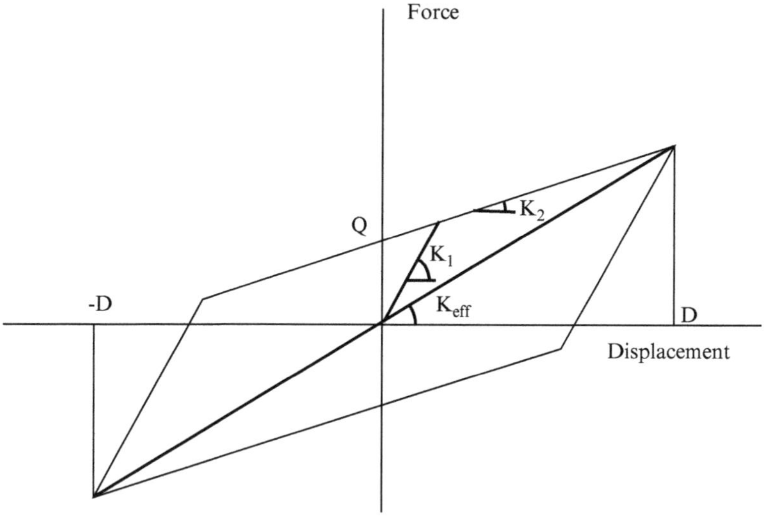

Fig. 2.3 Hysteresis parameters for elastomeric and lead rubber bearings (Naeim and Kelly 1999)

where, G = shear modulus of elasticity, A = pad area, h = pad thickness. Yielding is assumed to take place at 100% strain, and failure is assumed to take place at 150% strain.

The nonlinear hysteretic response of elastomeric bearings, lead-rubber bearings as well as Friction Pendulum Systems (FPS) can be essentially modelled using the bilinear relationship shown in Fig. 2.3. In practice, all isolation bearings can be modelled based on the three parameters K_1, K_2, and Q. The elastic stiffness K_1 is either estimated from available hysteresis loops from experiments or as a multiple of K_2 for LRB and FPS bearings. The characteristic strength Q is also characterized experimentally for rubber bearings, however, for LRBs it is given based on the yield stress in the lead and the area of the lead plug, while for FPS bearings it is given by the friction coefficient of the sliding surface and the load carried by the bearing (Naeim and Kelly 1999). SAP2000 (CSI 2007) provides a special biaxial isolator element that has coupled plasticity properties for two shear deformations, and linear effective stiffness properties for the remaining four deformations. The plasticity model is based on the hysteretic behaviour proposed by Park et al. (1986) and recommended for the analysis of base-isolated systems by Nagarajaiah et al. (1991). The accuracy of this element in SAP2000 was verified against experimental data. Element models for friction pendulum systems are discussed in Zayas and Low (1990) and adopted in SAP2000. Specialized elements for these bearings can be found in other standard structural analysis packages such as GT Strudl (2006) and are summarized in Buckle et al. (2006).

There are three common high-load multi-rotational bearing types that function in essentially the same manner: pot bearings, disc bearings, and spherical bearings (e.g. Fig. 2.4). These bearings are typically designed to allow superstructure rotation, or may also allow relative translation in one or more directions. The most common type of rotating bearing is the pot bearing while translation may be provided by

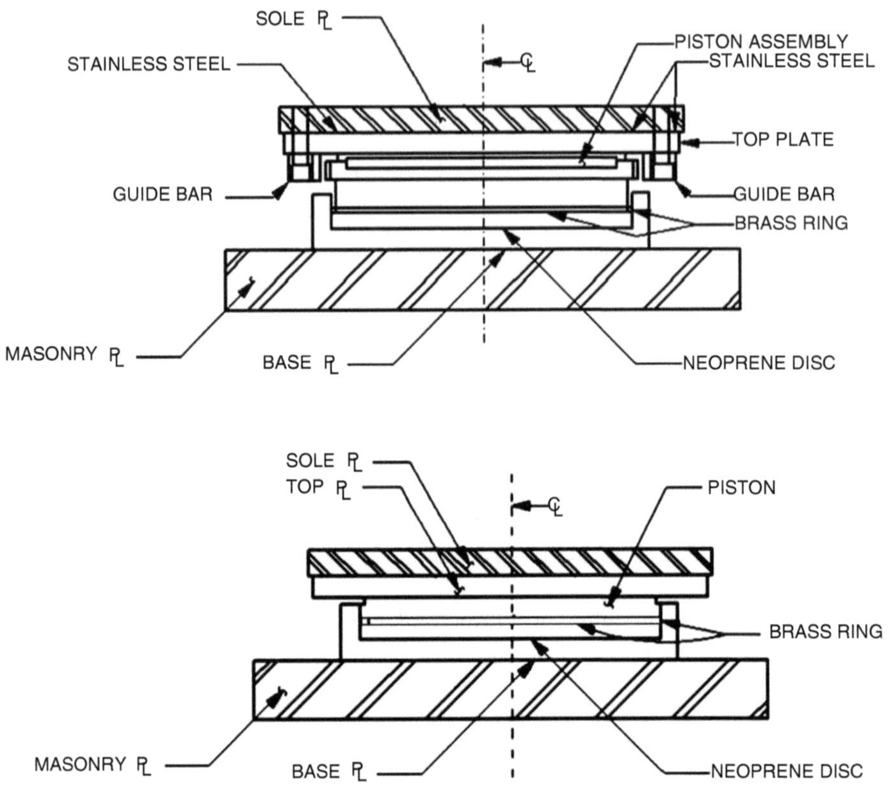

Fig. 2.4 Expansion and fixed multi-rotational bearings

elastomeric bearing pads, by polytetrafluoroethylene (PTFE) stainless steel sliders, or by rocker bearings. The force-deformation response of these types of bearings –if not restrained by design– can be characterized essentially with either elastic stiffness, or by Coulomb friction. It is noted that in the latter case, the nonlinear link element for FPS bearings in SAP2000 can be used with a zero (0) curvature, i.e. essentially flat surface. It is commonly accepted that PTFE bearings may be used as seismic isolators, but should only be used in conjunction with other devices that provide recentring.

2.3.4 Modelling of Shear Keys

Shear keys are commonly used in bridges with seat-type abutments to provide transverse support for the superstructure under service loads and earthquakes. They do not carry gravity loads, but in the event of an earthquake they are required to transfer the lateral reactions of the superstructure to the abutment or across movement joints. Sacrificial shear keys serve as structural fuses to control damage in abutments and the supporting piles under extreme transverse seismic loads. Sacrificial shear keys may be interior or exterior; exterior shear keys are usually

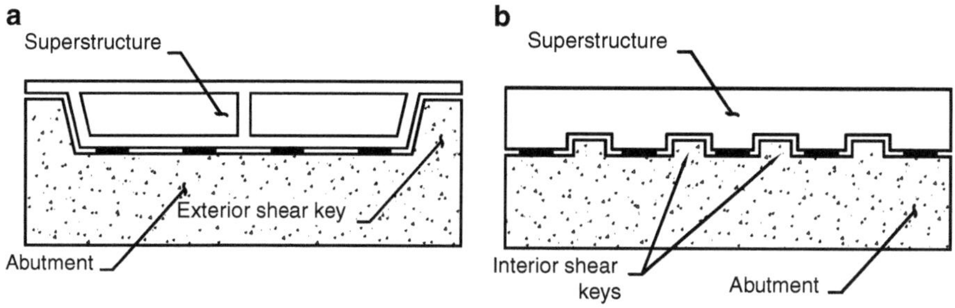

Fig. 2.5 Schematic view of exterior and interior shear keys

recommended for new construction because they are easy to inspect and repair. Interior shear keys are constructed within the abutment, and exterior shear keys are provided at the sides of the superstructure (Fig. 2.5).

Typically, for the design of highway bridges in seismic zones two distinct conditions are assumed (1) shear keys remain intact and (2) shear keys fail completely. These two conditions are intended to establish conservative bounds on the deformation and force response of other structural elements such as bent columns, piles, etc. (Caltrans 2006). In view of these preconditions, computer models of a given bridge are developed both with and without the shear keys. The former may be provided with proper elastic stiffness associated with the shear key design and geometry, or alternatively the transverse displacements at the shear key location are restrained. In order to identify the state of the shear key, force demand due to the earthquake is compared with the calculated capacity of the shear key. Capacity of shear keys is normally calculated based on a shear friction theory. The capacity may be also estimated using the strut and tie method (ACI 2008). Experimental studies of large-scale models of shear keys have shown that the mode of failure may be different than the shear friction method (Silva et al. 2003) for seismic demands below the capacity of the shear keys, they can be modelled by using a compression-only gap element. The stiffness after gap closure would be based on the cracked shearing stiffness of the shear key. For seismic demands beyond the capacity of the shear key, the computer models should ignore shear keys.

Recently, there has been considerable research on the failure of plain and reinforced concrete shear keys. Laboratory experiments conducted on the seismic performance of shear keys designed according to current seismic design criteria have indicated that there may be significant variability in the actual strength versus design values (Bozorgzadeh et al. 2006; Megally et al. 2001). These observations prompted the need to better model the complete nonlinear response characteristics for the seismic response assessment of bridges. Moreover, recent analytical studies have demonstrated that consideration of the nonlinear shear key response may be critical for the reliable and accurate assessment of the overall seismic response of highway bridges (Saiidi et al. 2001); particularly irregular bridges with large skew (Goel and Chopra 2008; Abdel-Mohti and Pekcan 2008). The latter study demonstrated that when bridges were subjected to large biaxial ground motion excitations, complete shear key failure –almost exclusively- preceded the unseating at the abutments of skew bridges. On the other hand, in bridges with some of the shear keys in near failure condition, the unseating may still be prevented. Clearly, it would only be possible to capture this response, if shear keys with realistic hysteretic properties are explicitly modelled.

Figure 2.6b shows an idealized force-deformation envelope for exterior shear keys. This idealized hysteretic model was demonstrated in terms of two models which correspond to the concrete and steel components. The above mentioned skew bridge studies followed the procedures presented by Megally et al. (2001) for the capacity determination of external shear keys. While the modelling of this unique hysteretic response behaviour may be accomplished relatively directly using analysis software such as Drain-2DX or Drain-3DX (Fig. 2.7), the same nonlinear external shear-key response can be modelled using a combination of link elements as shown in Fig. 2.6a using SAP2000, OpenSEES (Mckenna et al. 2000), or other similar software.

Experimental cyclic response of a shear-key (Megally et al. 2001) is compared to that of a key modelled as described above using SAP2000 in Fig. 2.8. Finally, it is noted that the abutment transverse shear key behaviour can be modelled using the elements depicted in Fig. 2.8. These elements can address the structural capacity of the shear key as well as contribution of passive resistance of the abutment embankment in the transverse direction as a function of relative displacement between bridge deck and abutment. The model may include gap between shear key and deck, and the limiting passive capacity of the embankment soil in the transverse direction.

2.4 Isolation and Energy Dissipation Devices

Several types of isolation systems are in use today and many new solutions are continuously being proposed and investigated (Skinner et al. 1993; Higashino and Okamoto 2006; Naeim and Kelly 1999). Most currently used isolation systems include: (i) lead rubber bearings (LRB) (Fig. 2.9a), (ii) high-damping or added-damping rubber bearings (HDRB/ADRB) (Fig. 2.9b), (iii) friction pendulum bearings (FPB) (Fig. 2.9c), (iv) Combinations of either low-damping rubber bearings (LDRB) or FPB with viscous dampers (VD), (v) Combinations of flat sliding bearings (FSB) (Fig. 2.9d) and LDRB, (vi) Combinations of FSB and steel yielding (SY) devices (Fig. 2.9e), (vii) Combinations of FSB, SMA (shape memory alloy)-based re-centring devices (Fig. 2.9f) and VD. Sliding bearings used in seismic isolation typically exploit the low friction between PTFE (Polytetrafluoroethylene or Teflon) pads in contact with lubricated polished stainless steel surfaces.

Basically four different idealized force-displacement models can be used to describe the cyclic behaviour of the aforementioned systems. The model parameters are identified in Fig. 2.9 and typical values are reported below. They have been derived from literature review and examination of product brochures of several seismic isolation manufactures.

The first model (see Fig. 2.9a) represents a visco-elastic behaviour. It can be used to describe the cyclic behaviour of HDRB and low damping rubber bearings (LDRB) (Taylor et al. 1992). The main difference between HDRB and LDRB

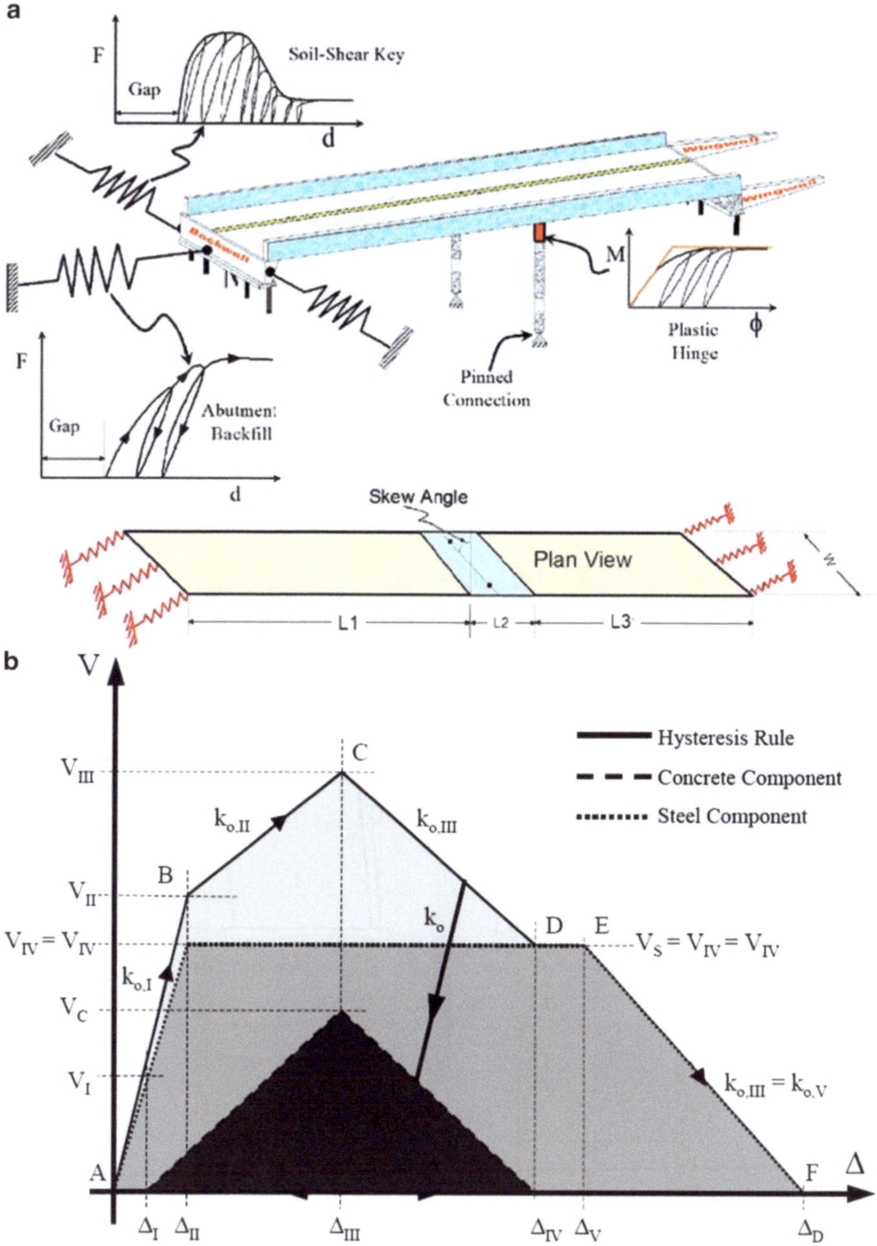

Fig. 2.6 (**a**) Components of a bridge model (Shamsabadi 2006). (**b**) Hysteresis rule for exterior shear keys (Megally et al. 2001)

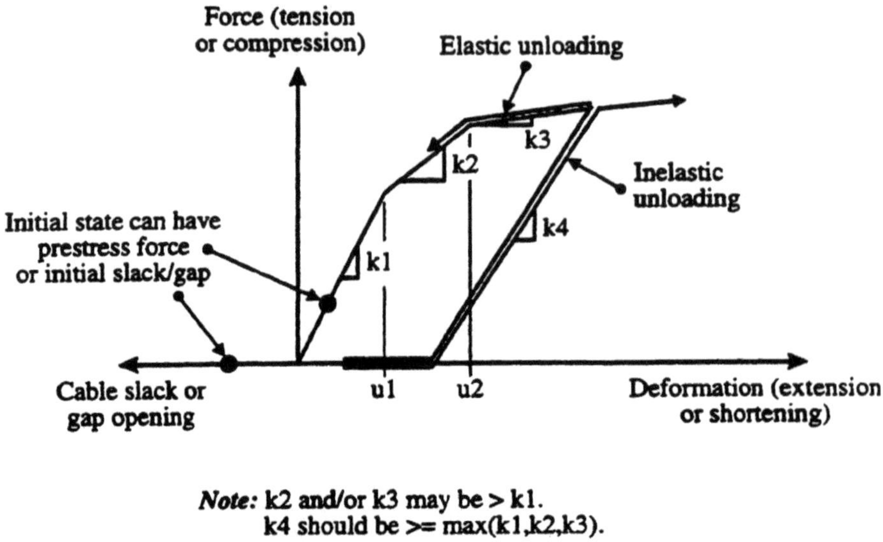

Fig. 2.7 Type 09 element in drain-2DX and drain-3DX

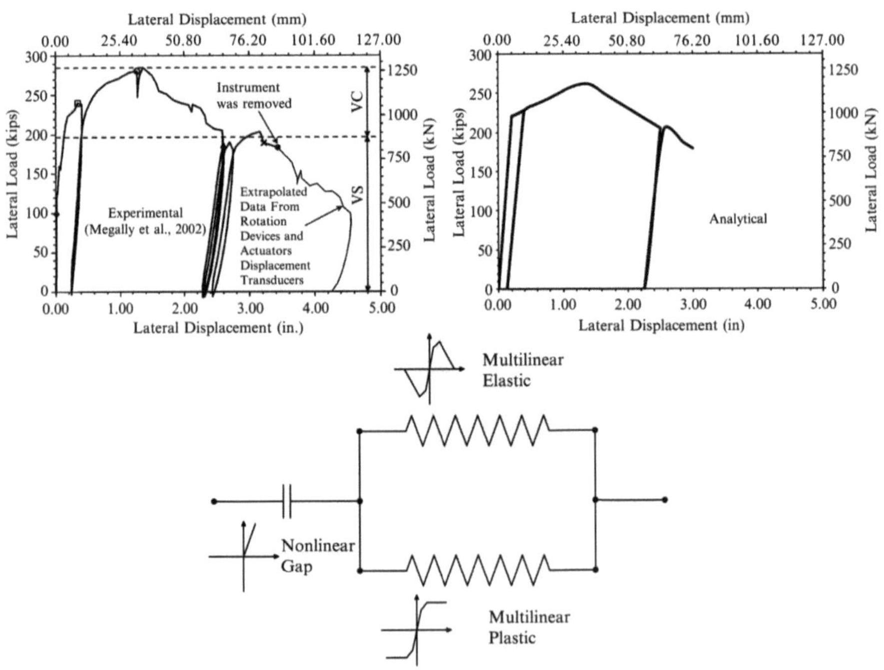

Fig. 2.8 Modelling of inelastic shear key force-deformation response

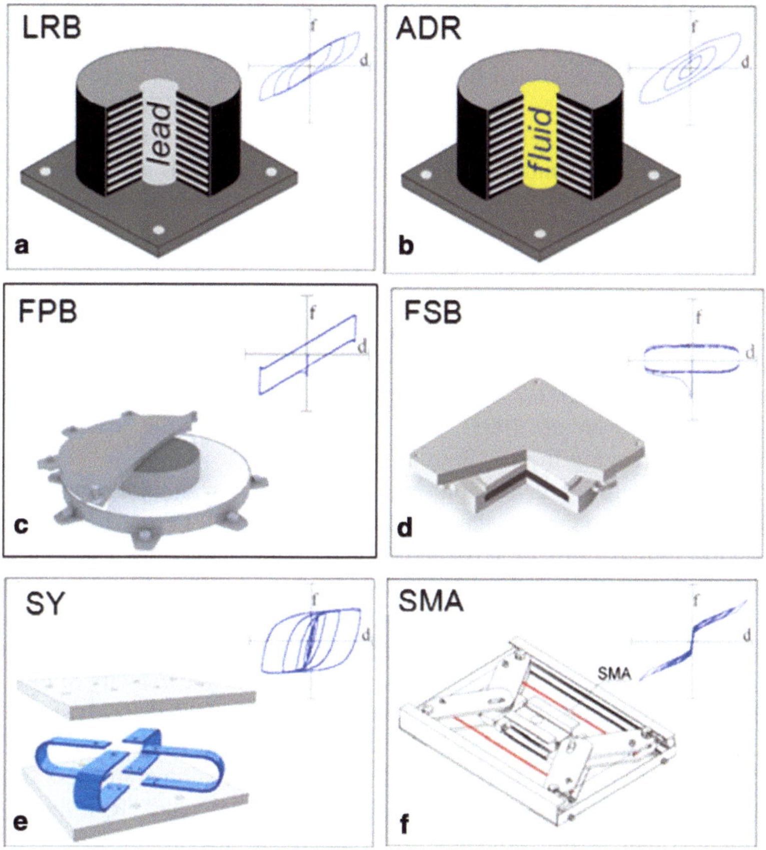

Fig. 2.9 Common isolation systems and associated force-displacement behaviours: (**a**) lead rubber bearings, (**b**) added-damping rubber bearings, (**c**) friction pendulum bearings, (**d**) flat sliding bearings, (**e**) steel yielding devices, (**f**) SMA-based re-centring devices

consists in the damping capacity of the elastomeric material. The elastomers (either natural or synthetic rubber) used in LDRB usually provide viscous damping ratios not greater than 5%. The rubber compounds used in HDRB, however, provide an effective damping which is around 15–18% at low (25–50%) shear strains, reducing to 12–15% for shear strains greater than 100% (Derham et al. 1985). Using the ADRB (Dolce et al. 2003), equivalent viscous damping ratios as high as 20–25% can be reached.

The second model (see Fig. 2.9b) represents a bilinear with hardening behaviour. It can be used to describe the cyclic behaviour of either SY devices or LRB (Skinner et al. 1993). The lead plug of LRB exhibits an initial shear modulus equal to approximately 130 MPa (compared to that of rubber which typically ranges from 0.4 to 1.2 MPa at 100% shear strain) and a yield shear strain of approximately 7.7% (Kelly 1992). The yield displacement of LRB (D_y in Fig. 2.9b) is, therefore, approximately equal to 7.7% H, H being the height of the device. Typical heights of LRB vary from 100 to 350 mm. As a result, yield displacements ranging from 7.7 to 26.7 mm are then expected for LRB. In practice, the cross section area of the lead plug is around 5–10% the gross section area of rubber (Skinner et al. 1993). As a consequence, post-yield stiffness ratios (r in Fig. 2.9b) of the order of 5–15% are expected. The maximum displacement capacity of LRB is governed by the allowable shear strain of rubber and by the global stability of the device under vertical load. Hence, it is governed by the total height of rubber and the cross section dimensions of the device (diameter D for circular bearings, side dimensions b_x and b_y for rectangular bearings). Typically, design displacements between 120 mm and 350 mm, corresponding to rubber shear strains of 100–120% and lead ductility ratios of 10–20, can be assumed for LRB, although much greater rubber shear strains (200–250%) and ductility ratios (20–40) can be cyclically sustained (Skinner et al. 1993). The aforementioned limits lead to effective damping ratios ranging from 15% to 25%.

The third model (Fig. 2.9c) represents a rigid-plastic with hardening behaviour, which can be used to capture the cyclic response of both FPS (Al-Hussaini et al. 1994) and combinations of FSB and LDRB. Sliding bearings used in seismic isolation typically exploit the low friction between PTFE pads in contact with lubricated polished stainless steel surfaces. The dynamic friction coefficient of PTFE- steel sliding bearings (m_{FR} in Fig. 2.9c) depends on a number of factors, e.g.: the sliding surface conditions, the bearing pressure, the velocity of movement and the air temperature. The friction coefficient of lubricated PTFE-steel sliding bearings normally varies between 2% and 5%, while increasing up to 10–12% for pure PTFE-steel surface (Dolce et al. 2005). The post-sliding stiffness of single concave FPB is defined as W/R_c (Al-Hussaini et al. 1994), where R_c is the effective radius of curvature of the sliding interface and W the supported weight. Similarly, for double concave FPB with equal friction on the two sliding interfaces, the post-sliding stiffness is defined as $W/(R_{c1} + R_{c2})$, R_{c1} and R_{c2} being the radii of curvature of the two sliding surfaces. In principle, there is no theoretical limit to the displacement capacity of FPB, provided that a device with the required dimensions can be manufactured. Actually, the horizontal displacement capacity of FPB is conditioned by the acceptability of the corresponding vertical displacement and residual horizontal displacement. Both are a function of the radius of curvature R_c. As a consequence, limitations to the ratio between the design isolation system displacement D_d and the radius of curvature R_c are necessary to limit vertical and residual displacements. Reasonable values of the ratio D_d/R_c are between $\mu_{FR,max}$ (e.g. 5% for lubricated interfaces) (Naeim and Kelly 1999) and 15%. FPB with radius of curvature, ranging from 1 to 10 m, are commercially available (Naeim and Kelly 1999). Considering the aforementioned parameters, effective damping ratios will be in the order of 10–20%. The idealised cyclic behaviour of SB + LDRB is very similar to that of LRB, except for the recentring capacity of the systems, which relies upon the shear stiffness of rubber (kr). The viscous damping of rubber ($\approx$5%) leads to greater effective damping ratios (hence higher non-linearity factors) compared to FPB.

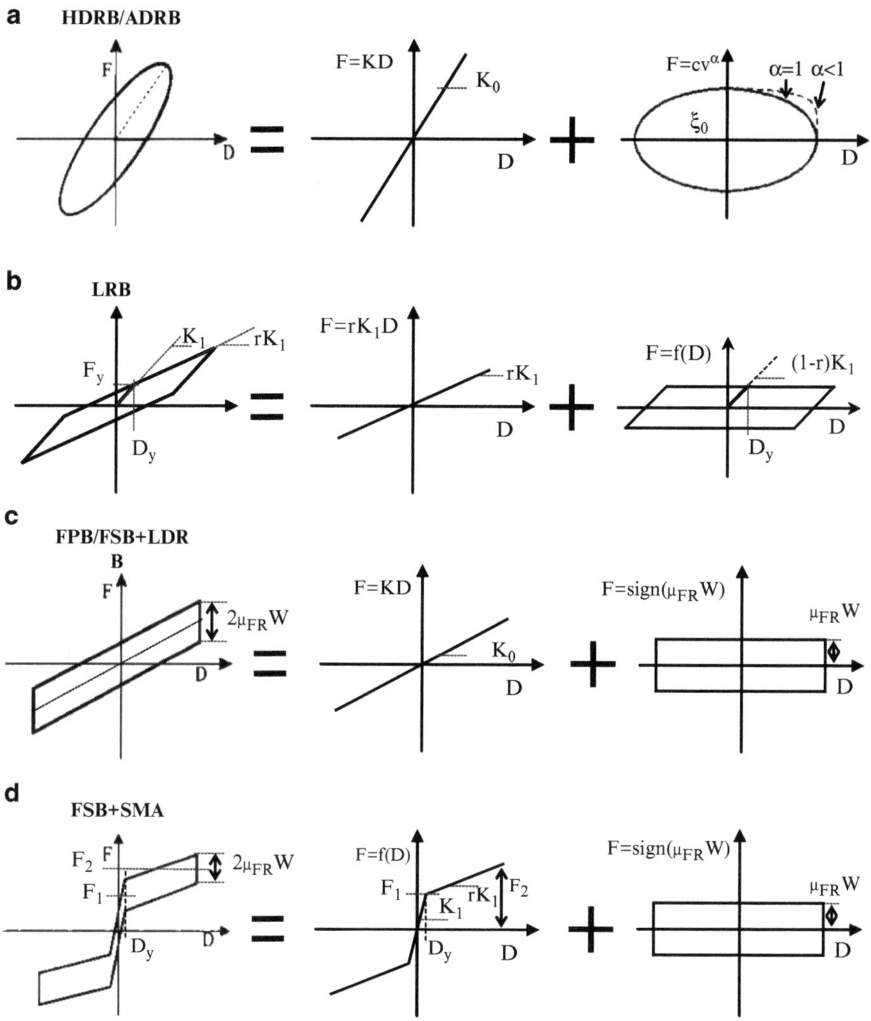

Fig. 2.10 Schematic force-displacement behaviour of the currently used isolation systems

As an alternative to LDRB, the SMA-based re-centring devices, designed, engineered and tested by Dolce et al. (2000) can be used. The SMA-based devices proposed by Dolce et al. (2000) exhibit a cyclic behaviour that can be schematized as bilinear elastic, although some energy is also dissipated by SMA, typically resulting in 3–5% damping. The fourth model (referred to as double flag shaped model) derives from the combination of a bilinear superelastic behaviour, modelling the typical force-displacement cycles of SMA-based recentring devices and a rigid-plastic behaviour, reproducing the schematic force-displacement cycles of FSB (see Fig. 2.10d).

Linear (a = 1) or nonlinear (1 < a < 0.2) viscous models (see right hand side of Fig. 2.9a) are used to take into account possible auxiliary viscous dampers (Constantinou et al. 1993).

The equivalent linear modelling of systems with nonlinear mechanical behaviour (i.e. LRB, SY, FPB, SB + LDRB, SB + SMA) is based on the definition of the effective stiffness and equivalent viscous damping, which accounts for the energy dissipated by the isolation systems during the seismic excitation through its viscous, frictional or hysteretic behaviour.

The effective stiffness (K_e) is defined as the secant stiffness to the design displacement D_d. The effective stiffness is related to the effective period of vibration (T_{IS}) of the isolated structure. Reference can be made to the following relationships:

$$K_e = (1.1 \div 0.9)\, K_0 \Rightarrow T_{is} = 2\pi \cdot \sqrt{\frac{W}{g \cdot K_0}} \quad \text{for HDRB/ADRI with } 50\% < g < 200 \tag{2.6}$$

$$K_e = \frac{K_1 + r \cdot K_1 \cdot (\mu - 1)}{\mu} \Rightarrow T_{is} = 2\pi \cdot \sqrt{\frac{W \cdot \mu}{g[K_1 + r \cdot K_1 \cdot (\mu - 1)]}} \quad \text{for LRB/SY} \tag{2.7}$$

$$K_e = \frac{W}{R_c} + \frac{\mu_{FR} \cdot W}{D_d} \Rightarrow T_{is} = 2\pi \cdot \sqrt{\frac{1}{g \cdot \left[\frac{1}{R_c} + \frac{\mu_{FR}}{D_d}\right]}} \quad \text{for FPB} \tag{2.8}$$

$$K_e = K_0 + \frac{\mu_{FR} \cdot W}{D_d} \Rightarrow T_{is} = 2\pi \cdot \sqrt{\frac{1}{g \cdot \left[\frac{K_0}{W} + \frac{\mu_{FR}}{D_d}\right]}} \quad \text{for SB + LDRB} \tag{2.9}$$

$$K_e = \frac{F_2^*}{D_d} + \frac{\mu_{FR} \cdot W}{D_d} \Rightarrow T_{is} = 2\pi \cdot \sqrt{\frac{1}{g \cdot \left[\frac{F_2^*}{W \cdot D_d} + \frac{\mu_{FR}}{D_d}\right]}} \quad \text{for SB + SMA} \tag{2.10}$$

where:

$$K_0 = G_0 \cdot (A_b/t_e)$$

G_0 is the shear modulus of the elastomer at 100% shear strain,
A_b is the effective horizontal area of the bearing,
t_e is the total thickness of the elastomer,
$\mu = D_d/D_y$ is the ductility ratio,
$r = (k_2/k_1)$ is the post-yield hardening ratio,

μ_{FR} is the friction coefficient of the sliding bearings,
R_c is the radius of curvature of FPB,
W is the total weight of the building (base floor included),
F_2^* is the force of SMA at the design displacement D_d,
g is the acceleration of gravity.

The equivalent viscous damping is computed based on either analytical-numerical formulations or semi-empirical relationships, as a function of the main mechanical parameters of the isolation system and the expected maximum displacement.

In first approximation, the equivalent viscous damping can be derived based on the well-known Jacobsen's equation:

$$\xi_{eq} = \frac{W_d}{4\pi \cdot W_s} = \frac{W_{viscous} + W_{hysteresis} + W_{friction}}{2\pi \cdot F_d \cdot D_d} \tag{2.11}$$

in which W_d is the total energy dissipated by the isolation system in the cycle of maximum amplitude (D_d), W_s is the strain energy stored at the design displacement D_d and F_d the corresponding force, equal to $K_e \cdot D_d$

The general expression of the equivalent damping ratio given in Eq. 2.11 can be specialized to each type of isolation system as follows:

$$\xi_{eq} = (1.1 \div 0.9)\,\xi_0 \text{ for HDRB/ADRI with } 50\% < g < 200\% \tag{2.12}$$

$$\xi_{eq} = \frac{2}{\pi} \cdot \frac{(r-1)\cdot(1-r)}{\mu \cdot [1 + r \cdot (\mu - 1)]} \quad \text{for LRB/SY} \tag{2.13}$$

$$\xi_{eq} = \frac{2}{\pi} \cdot \frac{\mu_{FR}}{\mu_{FR} + \frac{D_d}{R_c}} \quad \text{for FPB} \tag{2.14}$$

$$\xi_{eq} = \frac{2}{\pi} \cdot \frac{\mu_{FR}}{\mu_{FR} + \frac{K_0 \cdot D_d}{W}} \quad \text{for SB + LDRB} \tag{2.15}$$

$$\xi_{eq} = \frac{2}{\pi} \cdot \frac{\mu_{FR}}{\mu_{FR} + \frac{F_2^*}{W}} \quad \text{for SB + SMA} \tag{2.16}$$

For viscous-elastic isolation systems (i.e. HDRB/ADRI, see Figs. 2.9b and 2.10a), the equivalent damping ratio can be assumed practically constant and equal to the viscous damping at 100% shear strain. This implies a great flexibility in the choice of the design displacement of the isolation system (D_d), provided that the necessary damping ratio falls within suitable ranges.

For elasto-plastic isolation systems (i.e. LRB/SY, see Figs. 2.9a, e and 2.10b), x_{eq} depends on D_d through the ductility ratio. In principle, for a given design displacement D_d, it is possible to get a great variety of damping ratio values by properly selecting the yield displacement (D_y) and post-yield stiffness ratio (r) of the isolation system.

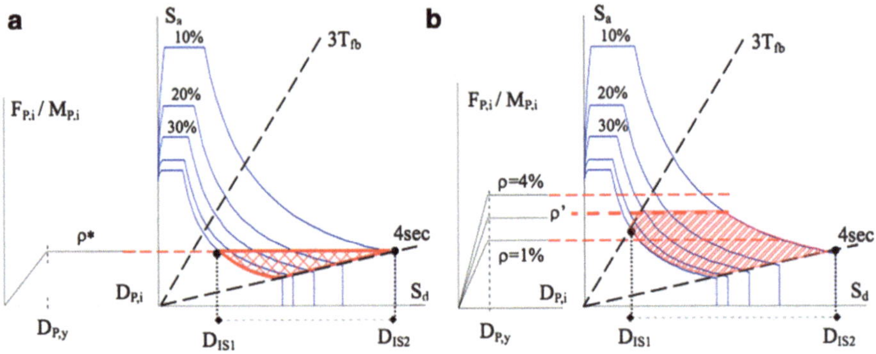

Fig. 2.11 Preliminary selection of isolation system type, design displacement (and pier reinforcement ratio) for the (**a**) retrofit of existing bridges and (**b**) design of new bridges

For friction-based isolation systems (i.e. FPB, FSB + LDRB, FSB + SMA, see Fig. 2.10c, d), x_{eq} depends on the friction coefficient (μ_{FR}), the design displacement of the isolation system (D_d) and the secant stiffness of the auxiliary recentring system/mechanism to the design displacement (i.e. W/R_c, K_0 and F_2^*/W for FPB, FSB + LDRB, FSB + SMA, respectively). This latter, however, is obtained only at the end of the design process. As a consequence, for FPS, FSB + LDRB, and FSB + SMA, an iterative design procedure is required, since input and output are mutually correlated. In other words, the design displacement of the isolation system (D_d) cannot be arbitrarily selected; otherwise the design procedure could fail to converge.

A preliminary selection of the isolation system type, design displacement, and (for new bridges) optimal pier reinforcement ratio is strongly recommended. This can be done using graphical tools similar to those reported in Fig. 2.11a, b, for existing and new bridges, respectively. The diagrams of Fig. 2.11 show a number of high-damping elastic spectra in the so-called ADRS (Acceleration-Displacement-Response-Spectra) format. Basically, each isolation system type is characterized by different damping levels. As a consequence, each type can be associated to a different group of response spectra. The dashed lines passing through the origin of the axis correspond to two limit values of the effective period of vibration of the bridge with seismic isolation, equal to $3T_{fb}$ (being T_{fb} the fundamental period of vibration of the bridge without seismic isolation) and 4 s, respectively. The interceptions of such radial lines with the response spectra corresponding to the lowest and highest values of damping ratio, define a preliminary range of possible isolation system design displacements. On the left hand side of each ADRS diagram of Fig. 2.11, the schematic displacement vs. acceleration relationship of the most critical (lowest shear/flexural strength) pier of the bridge in the direction of analysis is reported. For existing RC bridges (see Fig. 2.11a), for instance, the most critical pier is identified by a given reinforcement ratio (ρ*). For new RC bridges, instead, a suitable reinforcement ratio ($1\% < \rho' < 4\%$) can be initially assumed. Considering that, in presence of seismic isolation, piers are designed to remain elastic, only

Performance Points (PP's) that fall inside the painted background of the ADRS diagrams can be tentatively selected. Obviously, two further aspects affect the selection of the PP, i.e., the clearance of the joints and the isolation system displacement capacity. Once a suitable PP has been selected, the associated damping ratio (hence isolation system type), isolation period (hence effective stiffness) and (for new bridges) pier reinforcement ratio are identified. They can be used in the final bridge analysis and design.

2.5 Piers

2.5.1 Modelling Seismic Response of Columns in Reinforced Concrete Bridges

Bridges give the impression of being rather simple structures whose seismic response could be easily predicted. Therefore, for seismic design of the majority of bridges, simple models and simple methods are used in the practice. Design methods are predominantly elastic with only implicit consideration of the inelastic response. However, recent research has shown that certain structural characteristics may yield highly irregular and even unexpected response in the transverse direction of bridges (Paraskeva et al. 2006; Isakovic and Fischinger 2006 and Pinho et al. 2007). It has been realised that explicit inelastic seismic analysis is needed for unusual, complex bridges.

The dynamic response history analysis is one of the options, which could be employed to estimate the seismic response of RC bridges more realistically, taking into account their non-linear properties. However, a valid estimate of the seismic response obtained by this method depends on several critical issues such as an appropriate model of earthquake load, a sufficiently detailed numerical model of the bridge. In addition, the efficiency of computer software and its ability to analyze the bridge in reasonable time is critical. The modelling of the bridge piers is particularly critical and is the focus of this section.

The modern seismic design philosophy of standard (as opposed to unusual) bridges includes a consideration that damage of the bridge should be reduced to the flexural yielding of columns. There are several elements, which are suitable for modelling the non-linear behaviour of bridge columns; however, the knowledge about their applicability in practice is limited. In general, these elements could be classified as macro- or micro- elements. Macro-elements are different types of beam-column elements, where the non-linear behaviour is modelled using different hysteretic rules (force-displacement or moment-rotation relations), which attempt to capture overall member behaviour under cyclic loads. The basis of development of "hysteretic" – macro-models has been primarily experimental data. Therefore the parameters of hysteresis have clearly defined physical meaning. This makes macro-elements relatively easy to control. Since the hysteretic rules tend to represent the overall member behaviour, macro-models usually include substantially less elements than micro-models. This makes macro-models simple and more appropriate for complex dynamic non-linear analysis.

The second group of elements, micro-elements are, in general, plain (2D) or solid (3D) finite elements. Typically, the non-linear behaviour is modelled at the level of stress–strain relationships using predefined constitutive laws that are generally derived based on material test data. Compared to the macro-models, micro-models require substantially more calculations. This makes the already complex dynamic time-history analysis even more demanding on the analyst and the computer. Compared with the macro-models, the micro-models make the control of results and their analysis more complex and time-consuming.

Some macro-elements, e.g. fibre elements (discussed in subsequent sections) combine the properties of previously described types of elements. For example, fibre element is a truss type element, where the nonlinear behaviour is defined based on the stress–strain relationship of each fibre. It is more convenient to use macro-models when the global response of the entire bridge is of interest. The micro-models are more appropriate when the local responses of some details are studied, or when the structure includes complex details for which macro-elements are not available.

In view of the above considerations this section deals only with macro-elements. Three types of these elements: (a) beam-column element with lumped plasticity, (b) fibre element, and (c) MVL element, have been compared using two examples of four-span viaduct. This viaduct was originally investigated experimentally and analytically by Pinto et al. (1996).

Columns in substandard bridges designed prior to the development of modern seismic codes require special considerations under seismic loading. Very often these columns have insufficient transverse reinforcement, which can result in their shear failure and low ductility. To be able to model their response their shear strength should be properly assessed. In this section the shear strength of hollow box columns with relatively small amount of shear reinforcement, which was placed inside the longitudinal reinforcement, was analyzed experimentally and using the three analytical methods. Two analytical methods included in the Eurocode 8 standard (Part 2 and 3) as well as the method developed at the University of California (Priestley et al. 1996), San Diego were used.

Modelling bridge piers for seismic analysis and assessment can be done using different types of numerical tools. The following list is presented in increasing order of complexity and, perhaps, decreasing order of design applicability:

(i) Finite length plastic hinge discretization coupled with a hysteresis model.
(ii) Distributed flexibility based element with a multi-linear hysteresis model.
(iii) Fibre discretization using concrete and steel cyclic constitutive models.
(iv) FEM discretizations, either 2D or 3D, using a continuum damage cyclic model for concrete and the Menegotto-Pinto model for steel.

A general overview of modelling strategies is provided in the next sections complemented with few examples of application.

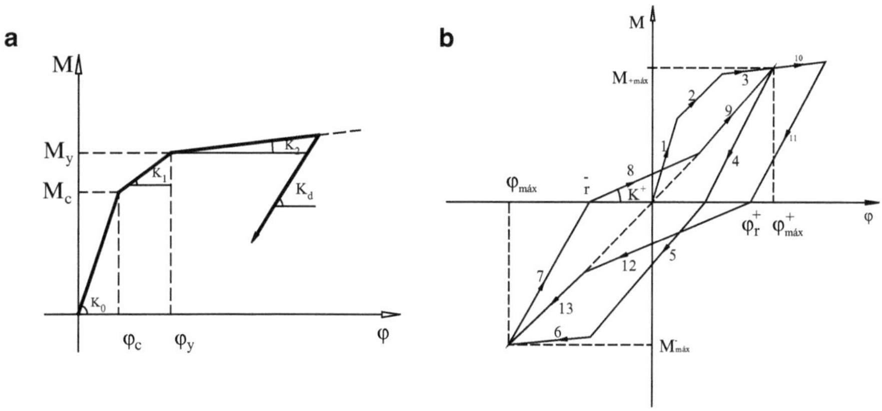

Fig. 2.12 Global section model. (**a**) Stiffness degradation; (**b**) pinching effect

2.5.2 Finite Length Plastic Hinge Model

This numerical model adopts structural beam-column elements with inelastic behaviour consisting of an elastic central zone and two end zones with inelastic response features. The non-linear material behaviour of such elements is therefore concentrated in their extremities, since those are the critical regions where yielding occurs over a short length, generally not greater than the cross section depth. In this model, cracking along the linear middle segment is not explicitly considered.

In the inelastic end-zone cross-sections, a global section non-linear model is adopted for simulating the reinforced concrete cyclic behaviour by means of a representative hysteresis model. The cyclic response is based on a skeleton curve (normally the monotonic response curve) coupled with a set of rules that control the characteristics of the loading-unloading-reloading cyclic reversals, typically addressing the pinching effect as well as the stiffness and strength degradation (Takeda et al. 1970; Saiidi 1982; Costa and Costa 1987; Duarte et al. 1990; CEB 1996).

Figures 2.12a, b show typical moment-curvature curves (M-φ) used to simulate the non-linear behaviour and the hysteretic energy dissipation characteristics for structural seismic response analysis, including the stiffness degradation and the pinching effect (CEB 1996). Although not shown in these figures, strength degradation is also possible to be included as fully addressed in Costa and Costa model (CEB 1996).

The response is initially linear elastic, turning into non-linear with the first cracks opening and proceeding to the post-yielding range when the tensile reinforcement yields. During cyclic loading the section stiffness reduces with increase of deformation that could be associated with strength degradation. This effect is controlled by the stiffness degradation parameter α, according to Eq. 2.1, wherein the reduced stiffness (K_d) is obtained from the elastic stiffness (K_0) affected by the

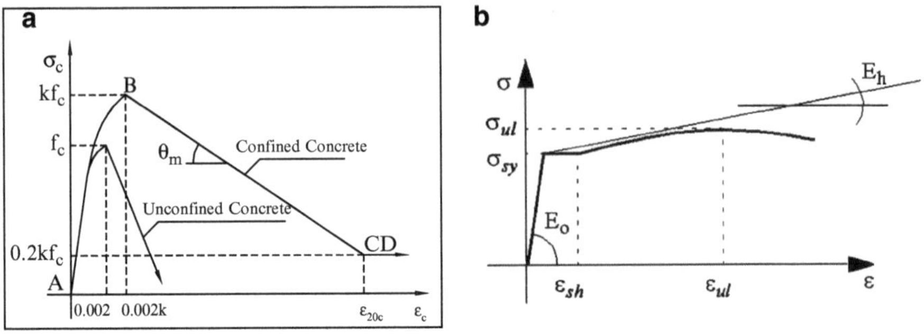

Fig. 2.13 Material behaviour models: (**a**) Concrete; (**b**) Steel

α-power function of the ratio of the maximum curvature ($\varphi_{màx}$) reached in a given cycle and the yielding curvature (φ_y).

$$K_d = K_0 \left| \varphi_y \middle/ \varphi_{máx} \right|^{\alpha} \tag{2.17}$$

The cyclic moment-curvature curve depicted in Fig. 2.12b illustrates the pinching effect, controlled by the parameter β according to Eq. 2.2 and can be observed in the reloading phases (lines 8/9 and 12/13), normally associated with crack closure.

$$K = m_{máx}/(\varphi_{máx} - \varphi_r)\left(\varphi_y/\phi_{máx}\right)^{\beta} \tag{2.18}$$

Strength degradation is usually associated with stiffness degradation, which is partially included in the model by imposing that the previous cycle maximum curvature during a given reloading stage is reached again for a target moment that is lower than the moment in the previous cycle. Details on this procedure are included in Costa and Costa (1987), which defines the strength degradation controlling parameter by a monotonically increasing function of the number of cycles and of the non-linear incursion magnitude expressed by the ratio of a given maximum curvature to the ultimate curvature (Wang and Shah 1987); such parameter is, therefore, analytically controlled by the curvature.

The monotonic (or skeleton) moment-curvature relationship is typically based on the initial concrete cracking, the tensile steel reinforcement yielding and its post-yielding behaviour, as well as the near peak concrete response. Based on the section geometric characteristics, the reinforcement detailing and the material characteristics, the skeleton curve can be obtained by fibre modelling using monotonic material constitutive laws as shown in Fig. 2.13a for concrete (both unconfined and confined according to Kent and Park (1971), Park et al. (1982), Kappos (1991)) and in Fig. 2.13b for steel. Alternatively, the monotonic moment-curvature curve can be defined as a multi-linear function of the characteristic points

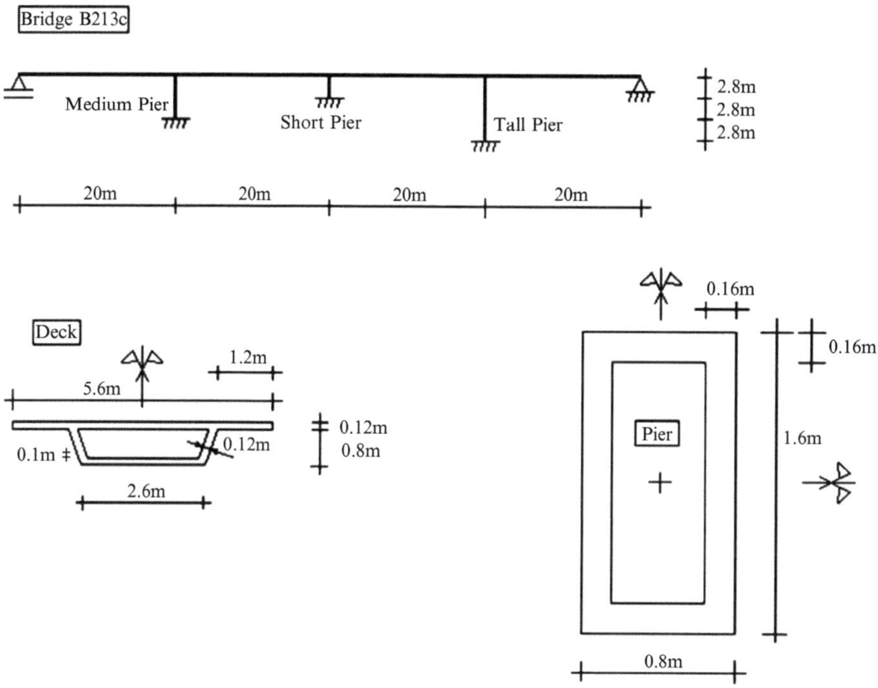

Fig. 2.14 B213c scaled bridge geometry

(cracking, yield and ultimate) which are explicitly determined using the procedure described in Arêde and Pinto (1996) for rectangular or T-shape RC sections without requiring any type of fibre discretization. In such case the "yield point" represents the effective yield point of the moment curvature relationship rather than the first yielding of reinforcement.

The finite length plastic hinge model was adopted for the simulation of experimental studies carried out on reduced scale models of bridge piers performed at the JRC-Ispra, Italy and reported in Guedes (1997). As part of a wider study within the framework of the PREC8 project, the scaled bridge B213c illustrated in Fig. 2.14 was considered, featuring a deck with four spans 20 m long and three hollow section piers with 5.6, 2.8 and 8.4 m high. Piers were fully-fixed at the base and pinned at the deck connection. The bridge was pseudo-dynamically tested with sub-structuring such that the deck was numerically simulated whereas the piers were tested at the reaction wall facility of the ELSA laboratory at Ispra by imposing top displacements and measuring the forces.

Figure 2.15 shows the comparison of experimental and analytical responses for the tall (Fig. 2.15a) and short (Fig. 2.15b) piers (Delgado et al. 2002). Concerning the numerical simulation, each pier was modelled by a single fixed length plastic hinge element with the measured lateral displacement history imposed on the top.

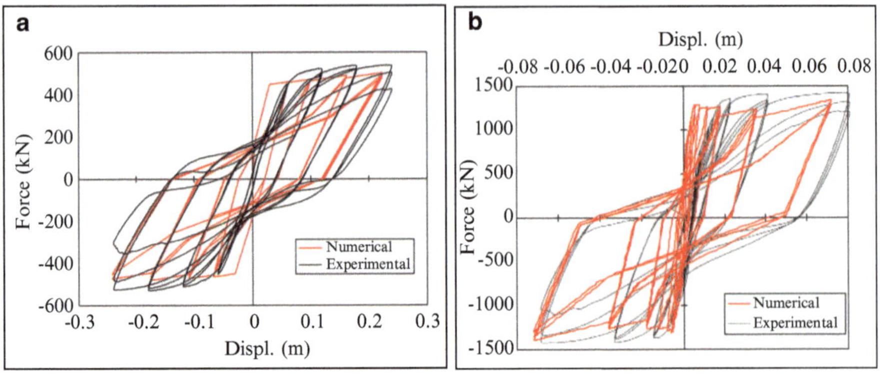

Fig. 2.15 Force-displacement curves for the B213c bridge piers: (**a**) tall and (**b**) short

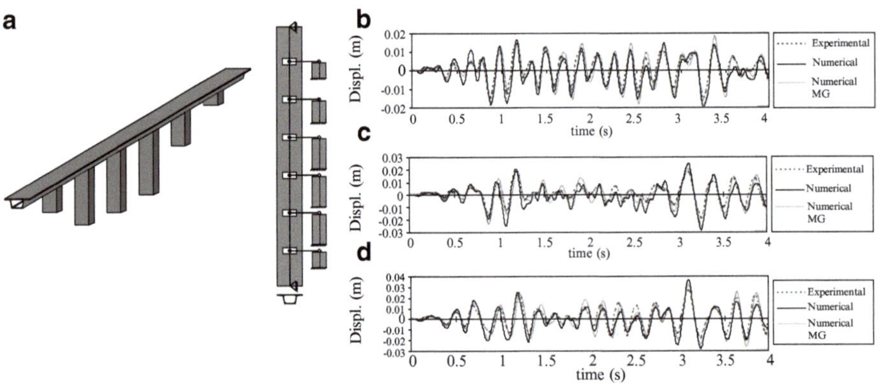

Fig. 2.16 Seismic analysis of B213c bridge. (**a**) Equivalent planar discretization and top-displacement time histories for (**b**) medium, (**c**) short and (**d**) tall piers (experimental and numerical)

The forces and the dissipated energy of these numerical results are reasonably close to the experimental results, thus suggesting good numerical model performance for simulating the inelastic behaviour of piers under cyclic loads with no significant computational cost. Therefore, this pier modelling was subsequently adopted for the seismic analysis of the bridge in the transverse directions using an equivalent planar discretization of the bridge as schematically shown in Fig. 2.16a, where the deck is assumed to be linear elastic and linked to the pier top sections by hinged, rigid and massless bars.

For the bridge numerical seismic analysis the same accelerograms were adopted as those used in the pseudo-dynamic experimental tests. Thus, non-linear dynamic response history analysis was performed and the results of pier top-displacements were compared against the experimental data. Moreover, to further confirm the model ability, these results were also compared with more refined simulations carried out by Guedes (1997) using fibre model simulation. These comparisons are illustrated in the displacement histories included in Fig. 2.16b, c and d for the medium, short and tall piers, respectively, where the experimental response is plotted in black dashed line; the numerical results are plotted in solid lines, black for those obtained by the fixed length plastic hinge model (time series labelled Numerical) and grey for the fibre model output (Numerical MG).

An overall observation of the above mentioned comparisons shows that the numerical simulations provide reasonably good agreement with the experimental response for the three piers, both in terms of frequency and of peak values. Despite some local deviations, globally the response is quite well captured and the fact that both plastic hinge and refined fibre modelling approaches yield similar results is itself an additional confirmation of the good balance between computational cost and result quality of the adopted plastic hinge model. Further details concerning both modelling and simulation strategies and assumptions can be found in Guedes (1997) and Delgado et al. (2002).

2.5.3 Distributed Flexibility Based Element Model

This global element model was developed (Arêde 1997) within the flexibility formulation framework for the analysis of reinforced concrete frame structures resorting to one-to-one discretization of structural members while duly accounting for progressive stiffness modifications along the member length. The algorithm was implemented in the general purpose finite element analysis code CAST3M, i.e. the former CASTEM 2000 (CEA 2003).

The development was pursued under the requirement that only one single element should be sufficient to describe the response of a given structural member, for which the classical displacement based formulation is not adequate because displacement shape functions change with member stiffness modifications caused by cracking and yielding. Therefore, the flexibility or force based formulation was adopted in line with previous works by Taucer et al. (1991) and Spacone et al. (1992), which proved to be particularly adequate for the required purpose since no displacement shape functions are required. In depth discussion can be found in Arêde (1997).

The general flexibility formulation resorts to a number of control sections where the behaviour is monitored and only makes use of force shape functions which have the main advantage of being correct regardless the element damaged state. The element flexibility matrix is obtained by integration of the flexibility distribution (known at the control sections) and an internal iterative scheme is adopted to obtain the element restoring forces corresponding to a given set of imposed displacements at the nodes as schematically shown in Fig. 2.16. Details on this scheme can be found in Arêde (1997).

While the above mentioned previous works dealt with fibre section modelling for flexibility and restoring force updating at predefined and fixed control sections, the global element model herein addressed resorts to a global section model

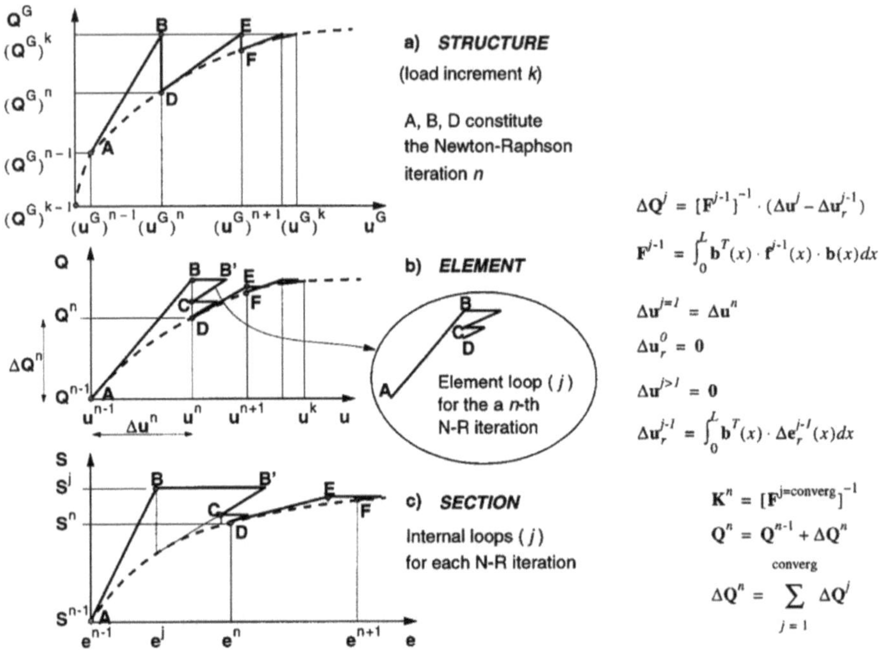

Fig. 2.17 State determination for force based (or flexibility) non-linear incremental scheme. Iterations for a given load increment: (**a**) at the global structure level, (**b**) at each element and (**c**) at the control section level

(based on moment-curvature behaviour laws) prescribed just at the element nodes and coupled with moving control sections that allow constantly monitoring the yielded, cracked and uncracked zones of the element. This model takes into account the section non-linear flexural behaviour for only one bending direction, by recourse to a global model based on moment-curvature (M-φ) trilinear skeleton curves as shown in Fig. 2.17a. The model is of Takeda type, requiring three parameters for the hysteretic behaviour control in each bending sense and allows simulating a wide variety of situations involving stiffness degradation, pinching effect and strength deterioration.

However, in order to make possible the control of cracking and yielding sections for each analysis step (respectively defined as the transition sections from cracked to uncracked zones and from yielded to cracked zones), some simplification is considered in the model, consisting in a sudden transition from the uncracked phase (branch 1 in Fig. 2.17b) to cracked behaviour (branches 2 and 3 in Fig. 2.17c) whenever any section reaches the cracking moment (M_C^+ or M_C^-) for the first time. Thus, both the uncracked and cracked stages are ruled by secant stiffness laws, greatly simplifying the control of sections matching such behaviour stages. After yielding, the behaviour rules are similar to those of the existing model with slight modifications.

It is noteworthy that, beyond the above mentioned simplification, any other type of Takeda type model is suitable for this purpose. In particular, the same section model (Costa and Costa 1987; Duarte et al. 1990; CEB 1996) as used for the finite

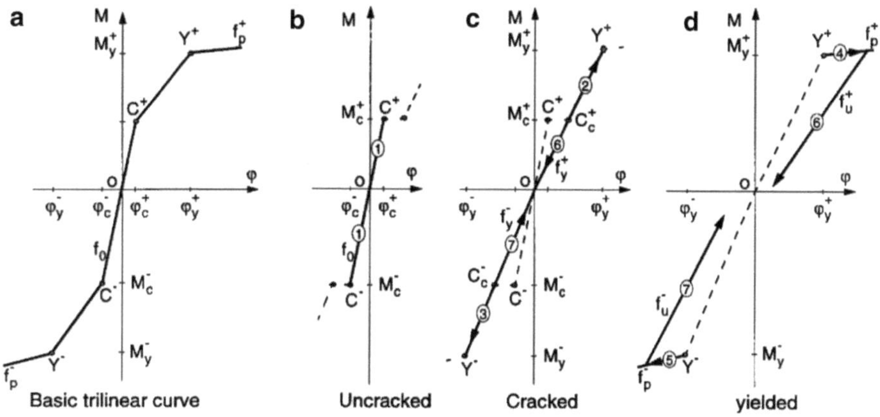

Fig. 2.18 Section model – primary curve and distinct behaviour phases

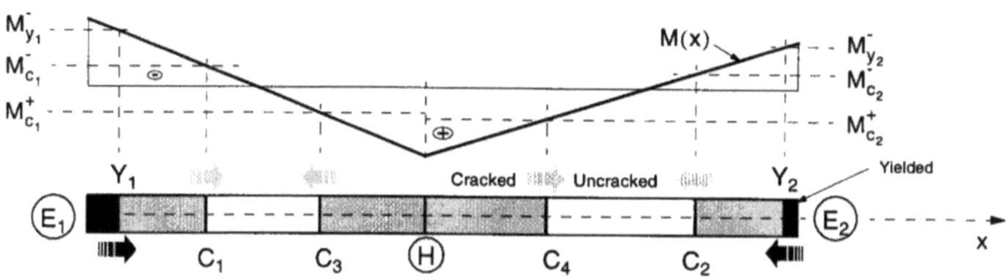

Fig. 2.19 General layout of yielding and cracking sections, and distinct behaviour zones

length plastic hinge formulation could also be adopted for this distributed flexibility global element model.

According to the above stated and as illustrated in Fig. 2.19, for each step the element is divided into yielded, cracked, and uncracked zones wherein adequate flexibility is assigned according to the respective section model. Control sections consist of fixed ones (the end sections, E1 and E2, and a central section H) and moving sections (the cracking ones, C1–C4, the yielding sections, Y1 and Y2, and, possibly, the null moment sections, O1 and O2). Sections E1 and E2 are fully controlled by the complete model, whereas the span section H is restricted to cracked behaviour because no yielding is assumed to take place along the central span zone. Cracking sections are controlled by the simplified model rules shown in Fig. 2.18b, c, depending on whether the uncracked or the cracked side is being considered, whereas sections Y1 and Y2 are controlled by the diagrams in Fig. 2.18c, d, respectively for their cracked and yielded sides. Null moment sections may also be activated in case they occur in cracked zones and are controlled by the rule of Fig. 2.18c.

Figure 2.19 refers to the case of a beam with a concentrated force applied in the span, but it is quite apparent that the same procedure is suitable for column analysis, with an even more simplified scheme for control section motion. This is evidenced in Fig. 2.19 which highlights different flexibility distributions along the element for a loading case of a beam (Fig. 2.19a) as shown in Fig. 2.20b and for an

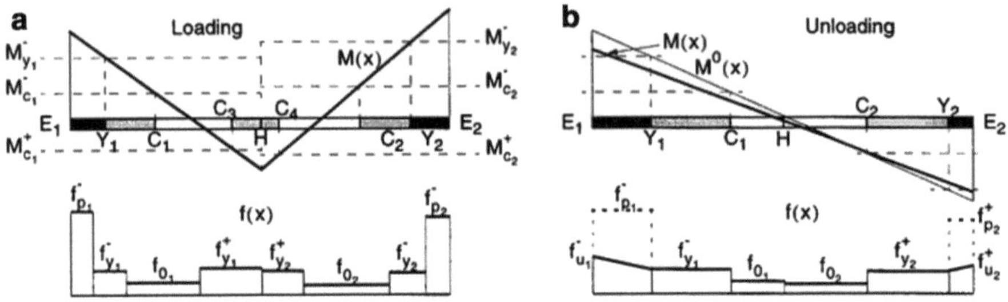

Fig. 2.20 Examples of flexibility distributions for (**a**) loading and (**b**) unloading cases

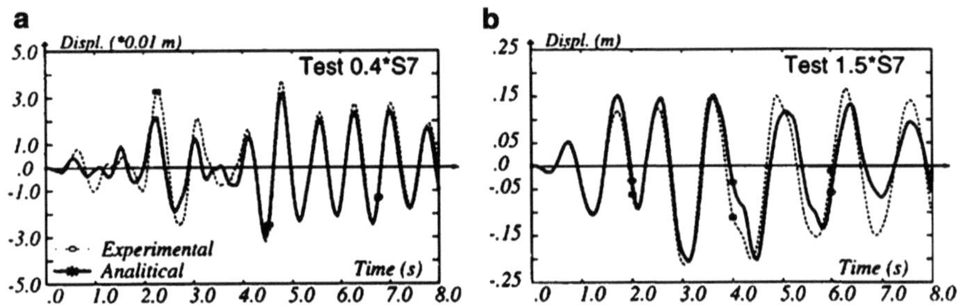

Fig. 2.21 Top displacements: dynamic simulation of (**a**) low and (**b**) high level tests

unloading situation (Fig. 2.19b) of an element loaded only by end-section moments. The latter case is that of major interest for bridge pier analysis since it applies both for cantilever or bent type piers. Thus, focusing essentially on that situation, Fig. 2.19b exemplifies the case wherein both yielded zones (the variable length plastic hinges) unload from the post-yielding branch (thus with constant flexibility as evidenced by the dashed line in the f(x) flexibility diagram) and for which a linear variation is assumed between the end-section and the yielding section flexibilities. Specific details of model development and implementation can be found in Arêde (1997).

Because the model was initially used for building analysis, no specific studies were carried out on the seismic response analysis of bridge piers. Nevertheless the model was investigated in Arêde (1997) for a four-story building and satisfactory results were obtained (Figs. 2.20 and 2.21).

2.5.3.1 Fibre Model

The fibre model is widely available beam-column element model (Taucer et al. 1991; Guedes et al. 1994; Guedes 1997; SeismoStruct 2005), which can be regarded as a step further in the refinement of standard beam models. In fibre modelling, the sectional stress–strain state of the elements is obtained through the integration of

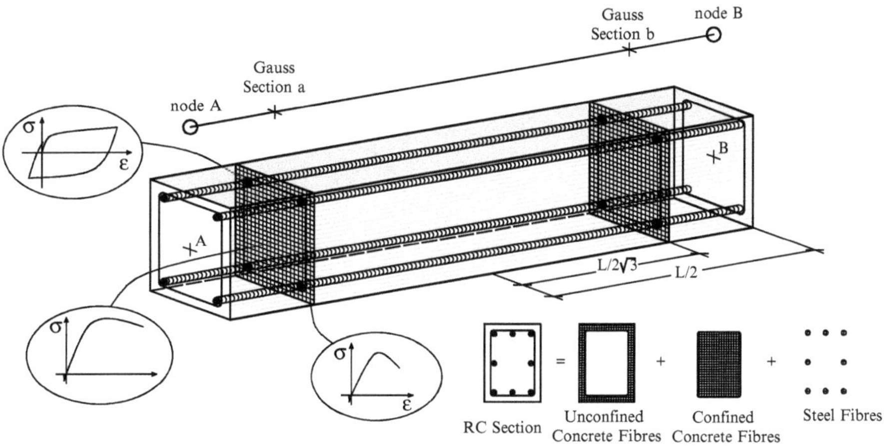

Fig. 2.22 Discretization of a typical reinforced concrete cross-section

the nonlinear uniaxial stress–strain response of the individual fibres in which the section is subdivided, distinguishing steel, confined and unconfined concrete, as illustrated in Fig. 2.22. The adopted stiffness-based element cubic formulation then allows both the representation of the spread of inelasticity along the member length as well as the implicit incorporation of interaction between axial force and transverse deformation of the element. The use of a sufficient number of elements per structural member permits the reproduction of plastic hinge, typical of members subjected to high levels of material inelasticity. The spread of inelasticity across the section and along the member length is thus achieved without requiring expertise calibration of any lumped plasticity element.

Structural members are represented by means of frame elements, with finite length and assigned cross-sections. Structural and non-structural inertia mass may also be introduced, in either lumped or distributed fashion, whilst joint/link elements, defined as spring-type elements joining coincident locations, can be used to model discontinuous connections. By means of such element types, a number of different element classes (columns, beams, walls, beam-column joints, etc.), non-structural components (energy dissipating devices, inertia masses, etc.) and different boundary conditions (flexible foundations, seismic isolation or structural gapping and pounding) can be represented. The fibre-discretization renders possible a realistic modelling of the different materials, and their distribution, that make up the cross-section of a given member. The material models may feature different levels of accuracy and complexity; the bilinear, the Menegotto-Pinto (1973) and the Monti-Nuti (1992) models are among the most used models for steel, whilst concrete may be characterized by tri-linear, nonlinear with constant or variable confinement constitutive laws (e.g. Scott et al. 1982). Many other material constitutive laws are available in the literature. The optimal number of fibres representing the cross section should be determined based on a series of batch tests with different mesh sizes (Sadrossadat-Zadeh and Saiidi 2007).

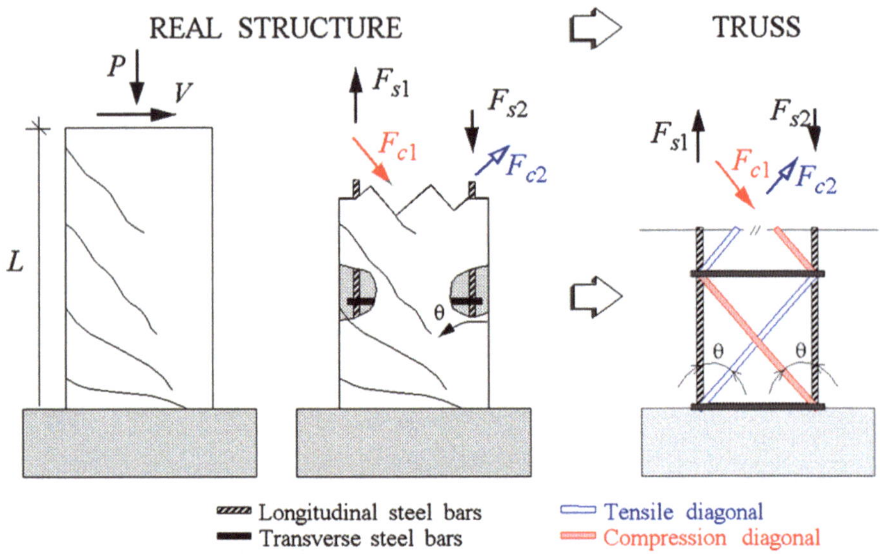

Fig. 2.23 Truss analogy for a cantilever beam under significant shear forces

The fibre model herein reported is implemented in the CAST3M software package (Guedes et al. 1994; CEA 2003) using a 3D Timoshenko type beam element so that interactions between axial and shear forces and bending moments can be considered. Each 3D element, with two nodes and six degrees-of-freedom per node, is divided into other longitudinal elements (the fibres) that react to both axial and shear displacements according to suitable constitutive laws. Each fibre is defined by a four or three node 2D element in the transverse section. Along the longitudinal (element) axis direction, the behaviour of the 3D elements is integrated using one Gauss point per element and the model is formulated for a longitudinal axis eccentric to the element gravity axis. Details on the element model formulation can be found in several publications (Guedes et al. 1994; Guedes 1997).

Experimental evidence concerning the cyclic response of shear dominated elements, such as squat bridge piers, show that non-linear shear behaviour must be considered to achieve accurate numerical response simulations. Therefore, the above mentioned classic fibre model was coupled with a strut-and-tie model to simulate the non-linear shear behaviour by adopting the procedure presented in Garstka et al. (1993) and Guedes (1997) as described below.

Classic fibre models assume linear elastic behaviour for shear stress components. This approach is acceptable when shear deformations are very small, but needs to be augmented in analyzing short bridge columns by adding a nonlinear shear element. The shear element uses the truss analogy (Fig. 2.23) for a cantilever

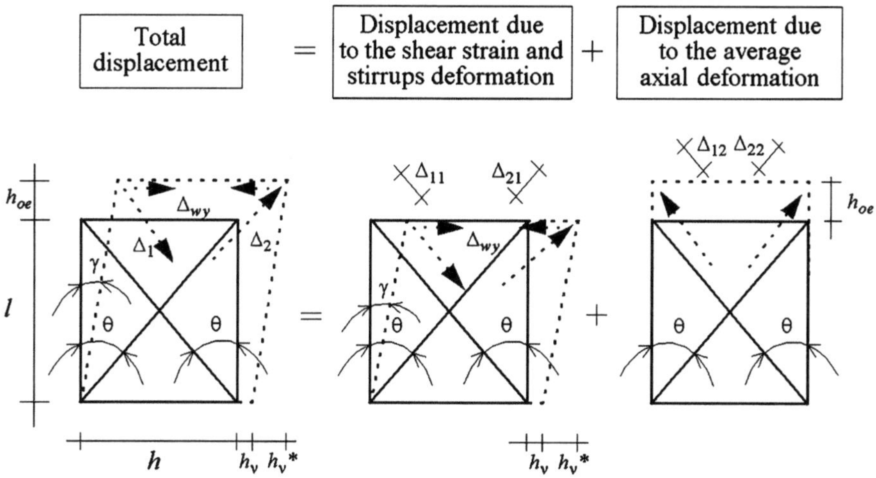

Fig. 2.24 Compatibility of displacements in the equivalent truss analogy

beam under shear forces. Thus, the non-linear shear model herein described is based on the assumption that shear strength is provided by the "equivalent" truss made of transverse and longitudinal steel ties and diagonal concrete struts that are formed due to diagonal cracking.

The model relies on displacement compatibility and force equilibrium in the cross-section. Other parameters are the constitutive laws of the materials and the shear cracking angle θ that defines the orientation of the concrete struts.

The displacement compatibility in the transverse direction is established by analysing the deformed truss shown in Fig. 2.24 subjected to axial force P, transverse force V and bending moment M. Two concrete diagonals are considered in the formulation for each loading sense: one for compressive and another for tensile forces. Detailed formulation is reported in Guedes (1997).

The non-linear shear model was adopted to simulate the response of a squat bridge pier tested at the ELSA laboratory in Ispra, Italy (Pinto et al. 1996). Both linear and the non-linear shear models were included in the study and the results were compared with the experimental data.

The pier was 2.8 m high and corresponded to a 1:2.5 scale specimen where a vertical force equal to 1.72MN was applied on the top to simulate the dead load of the deck. The mechanical properties of materials used in the tests corresponded to C25/30 concrete grade and B500 Tempcore steel (Fig. 2.25).

The top force-displacement diagrams are shown in Fig. 2.26, where comparison between experimental and numerical results is made.

Figures 2.24a, b show that the non-linear shear model reproduces quite well the narrowing and lengthening of the hysteretic force-displacement curve but the linear shear model overestimated the dissipated energy.

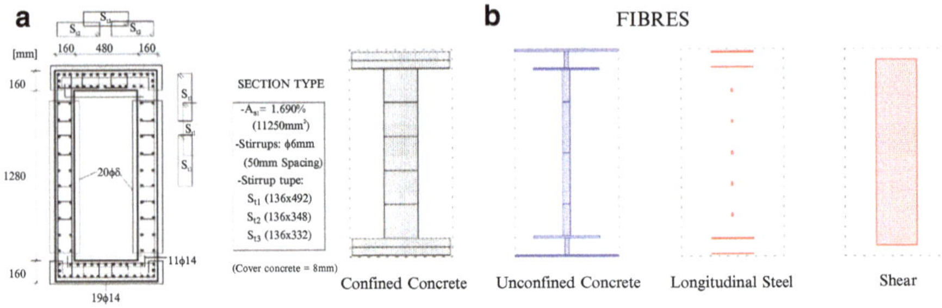

Fig. 2.25 Model validation. (**a**) Pier cross-section and (**b**) Corresponding fibre discretization

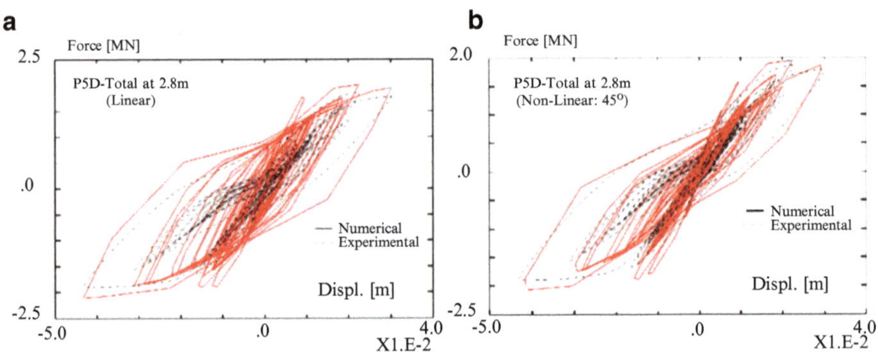

Fig. 2.26 Experimental *vs.* numerical top force-displacement results. (**a**) Linear shear model and (**b**) non-linear shear model (Guedes 1997)

2.5.4 Two and Three-Dimensional FEM Discretizations

In the framework of refined modelling of bridge piers, 2D or 3D FEM discretizations can be adopted wherein both concrete and steel are simulated using suitable constitutive laws, the former with plane or solid finite elements and the latter with uniaxial truss elements. Within this contribution, the behaviour of steel bar elements is ruled by the Giuffré-Menegotto-Pinto model (Menegotto and Pinto 1973), whilst for concrete a Continuum Damage Mechanics cyclic model (Faria et al. 1998, 2002) is adopted. Besides other stand-alone computer codes developed by the model author and collaborators, the model is implemented in the CAST3M package and has already proved to be suitable for seismic behaviour analysis of RC hollow bridge piers (Faria et al. 2004) and (Delgado et al. 2007).

The Continuum Damage Mechanics model adopted for concrete is cast with a strain-driven suitable scheme for simulating the degradation occurring under tension and compression (Faria et al. 1998) (Fig. 2.27).

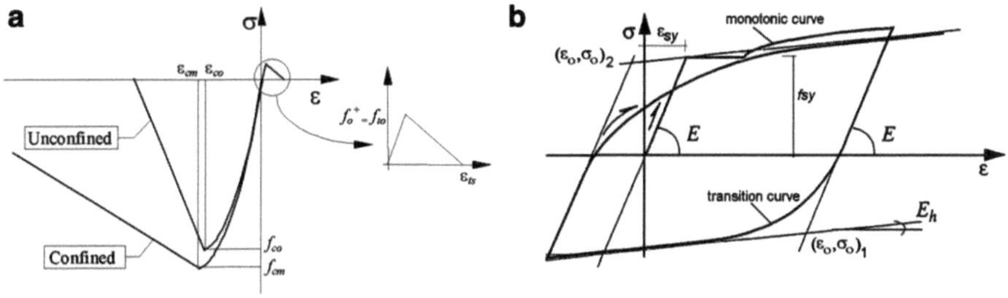

Fig. 2.27 Constitutive models' evolution under 1D loading. (**a**) Concrete behaviour and (**b**) steel cyclic model (Faria et al. 2004)

This model has been extensively used for seismic or cyclic response studies of RC walls (Faria et al. 2002) and bridge pier scaled models as reported in Faria et al. (2004), Delgado et al. (2009), Arêde et al. (2009). The conclusions generally point to good agreement between numerical and experimental results. The cyclic global force- displacement response curves were quite well estimated by the numerical model which also succeeded in adequately simulating energy dissipation, the strut-and-tie strength mechanisms experimentally observed on shear dominated structural members, the occurrence of localized or smeared cracks, as well as the formation of plastic hinges.

2.5.5 *Example 1 on Fiber Model Application*

The study of three numerical models suitable for modelling the flexural response is presented in this section. Simple beam-column element model with lumped plasticity and fiber beam-column element model were first investigated. Then the validity of the MVL (multiple-vertical-line), also known as multi-spring element (Saiidi et al. 1989 and Jiang and Saiidi 1990) element was also tested. The MVL element is a combination of the previous two types. The element consists of several springs, which are rigidly connected at the top and the bottom of the element. The cyclic response of each spring is controlled by hysteretic rules.

A large-scale specimen (1:2.5) of typical viaduct was analyzed in the transverse direction. Full-scale structures consisted of a 200-m deck and three single column bents (Fig. 2.28). The deck was pinned at the abutments. When modelling a viaduct, the superstructure was assumed to be elastic. Abutments were modelled as infinitely rigid. Columns were pinned at the level of the superstructure and fixed to the footings. Seismic load was defined with the generated earthquake record, used in the experimental studies.

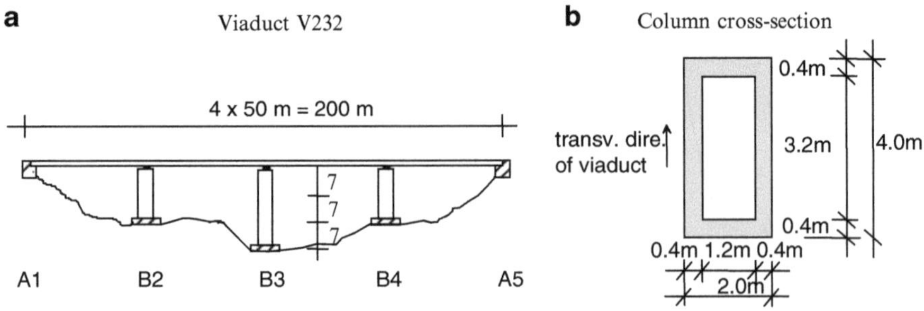

Fig. 2.28 Full-scale structure

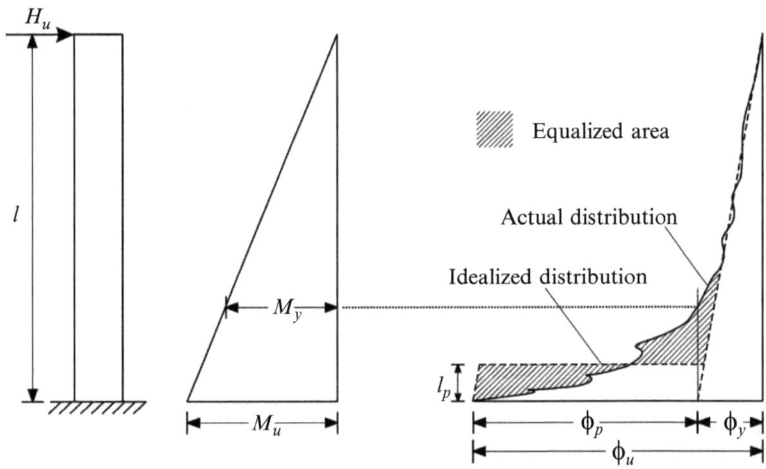

Fig. 2.29 Equivalent column and the definition of the equivalent plastic hinge length based on the idealized curvature distribution

2.5.5.1 Beam-Column Reinforced Concrete Element with Lumped Plasticity

In this model part of the element between the node and the point of contraflexure can be represented by an equivalent column (Fig. 2.29).

The drift ratio is then defined as the top displacement divided by the height of the column. The displacement is obtained by the double integration of the curvature along the height of the column. Typically for macro elements a number of idealizations have been used in the application. Linear distribution of bending moment and idealized curvature distribution are assumed. Plastic curvature is then considered to be constant over the equivalent plastic hinge length, which has been empirically determined (e.g. Kanaan and Powel 1973).

Although the element appears to be crude it is a realistic representation of structural members in which plastic hinging occurs at the ends. An application for hollow-box columns in a highway viaduct, described in previous sections is

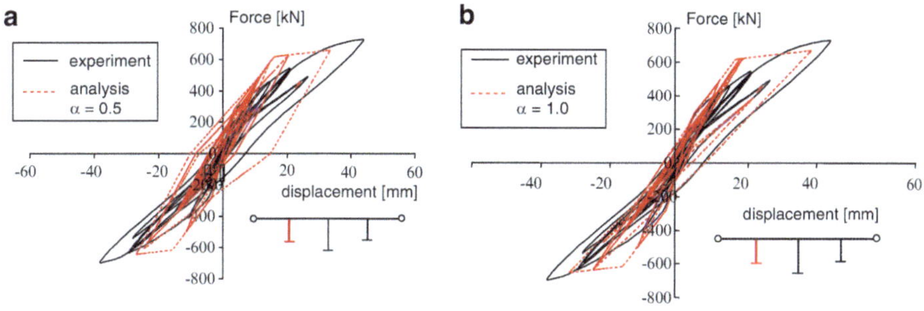

Fig. 2.30 Shear force-displacement diagram for the initial model ($\alpha = 0.5$) and model using $\alpha = 1.0$ (design earthquake)

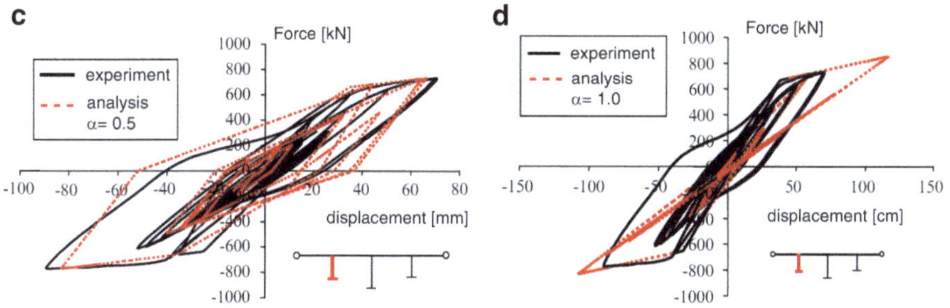

Fig. 2.31 Shear force-displacement diagram for the initial model ($\alpha = 0.5$) and model using $\alpha = 1.0$ (high-level earthquake)

briefly presented below. The beam-column macro element used in this study was incorporated into DRAIN-2D program (Kanaan and Powel 1973) at the University of Ljubljana (Fajfar and Fischinger 1987). In this element tri-linear Takeda hysteretic rules control the response of the rotational springs. In the initial model no "tuning" of the element parameters was done. All properties (including hardening parameter) were calculated from first principles. The common average value of the unloading parameter in the Takeda model ($\alpha = 0.5$), which determines the rate of the unloading stiffness deterioration, was used.

The correlation between the analytical and experimental displacement response histories, obtained by the initial model, was good and so was the modelling of the predominantly flexural hysteretic behaviour of the tall central column (Isakovic and Fischinger 1998). However, in the case of the design earthquake, the initial model underestimated the actual stiffness degradation on the unloading branch for the short column (Fig. 2.30). To account for higher stiffness degradation, unloading parameter $\alpha = 1.0$ should be used in the modified model. This change improved the calculated response in the case of the design earthquake, but not in the case of the stronger earthquake (Fig. 2.31), indicating that different α values should be used for different levels of response.

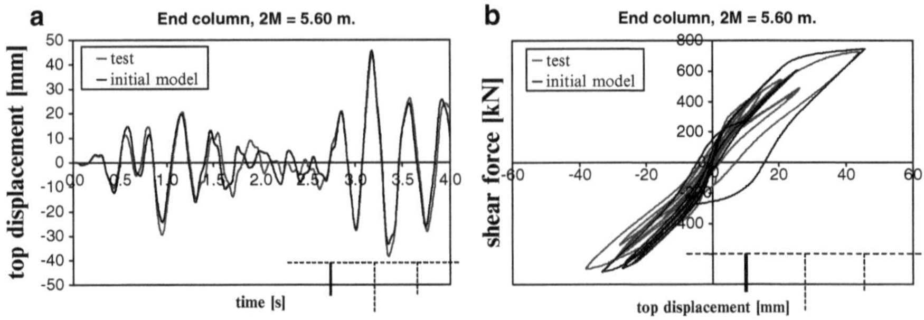

Fig. 2.32 Displacement history and shear force-displacement diagram (initial model)

These results are quite typical for the application of the macro elements. We can conclude that the overall results have been quite good in all cases. However, it is unrealistic to expect and claim that such, empirically based models can capture all the details of the response in the cases for which they were not calibrated. Nevertheless, even more refined models, like the stress–strain monitoring fibre element cannot guarantee good (or even at least acceptable) results.

2.5.5.2 Fibre Element

Among several possibilities in program OpenSees (Mazzoni et al. 2003) the force based non-linear beam-column element with distributed plasticity was used since it was found to be more efficient than standard displacement-based fiber element. Model of the column can consist of elements, which are further divided into smaller segments with different cross-sections. The correlation between analytical displacement time history response, obtained by the initial model (with typical values of characteristic parameters), and experimentally obtained displacements was good (Fig. 2.30). Similar to the initial beam column element with lumped plasticity, this model failed to estimate actual stiffness degradation on the unloading branch (Fig. 2.30b). Several improvements were necessary to obtain better results (Fig. 2.31). Model of concrete and model of steel as well as the number of integration points were changed. For example: the strength of the concrete in tension was taken into account and, instead of the bilinear stress–strain relationship, the Giufré-Menegotto-Pinto (Maekawa et al. 2003) model for steel was used, etc. (Figs. 2.32 and 2.33).

2.5.5.3 MVL (Multiple-Vertical-Line) Element

MVL element includes also a horizontal spring to model shear behaviour (Fig. 2.34). This element was originally proposed by Japanese researchers (Kabeyasawa et al. 1983) and later modified by Vulcano et al. 1989 and Fischinger et al. (1992).

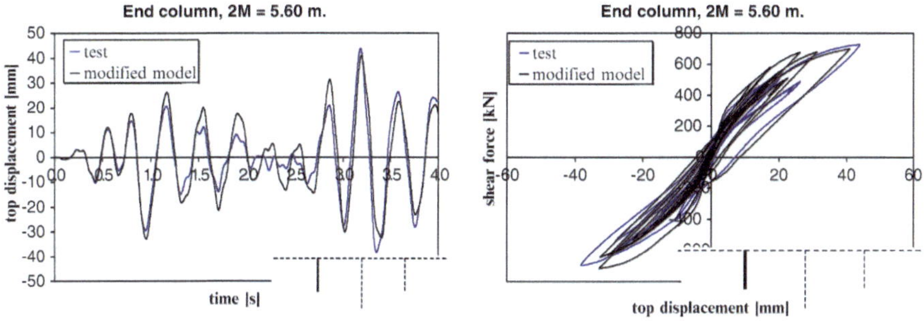

Fig. 2.33 Displacement history and shear force-displacement diagram (modified model)

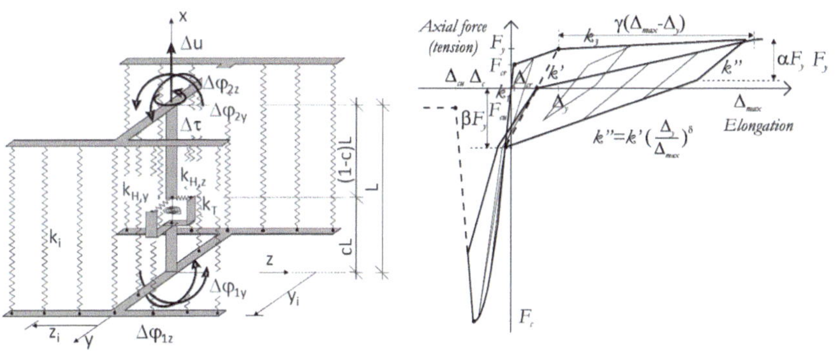

Fig. 2.34 Multiple-Vertical-Line-EleMent (MVLEM) and hysteretic rules of vertical springs

All these versions of the element were for the analysis of the unidirectional response only. Another version of the element has been developed (Fischinger et al. 2004) for bi-directional analysis of structures.

Each column of the viaduct was modelled with nine MVL elements. The displacement time-history obtained with the initial model (using standard parameters: $\alpha = 1.0$, $\beta = 1.5$, $\gamma = 1.05$, $\delta = 0.50$), was reasonable (Fig. 2.35). The estimation of the stiffness degradation on the unloading branch was better than that obtained with the previous two elements. Therefore, standard parameters were not changed.

2.5.5.4 Comparison of the Models

It can be concluded, that all three models are suitable for modelling the global behaviour of viaduct columns. All the initial models (using standard values of parameters) estimated the maximum displacements, as well as the maximum forces, quite well. Some discrepancy with the experiment was detected mostly during the unloading phase.

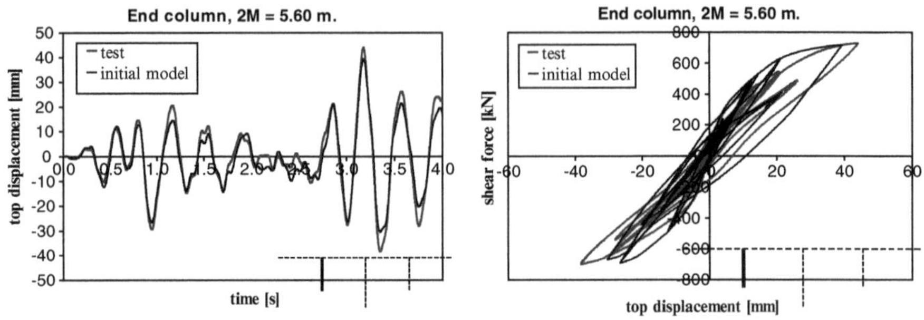

Fig. 2.35 Displacement time-history and shear force-displacement diagram (initial model)

Table 2.2 Advantages and limitations of the presented elements

Type of element	Advantages	Limitations
Beam-column element with lumped plasticity	Simple model with small number of elements (often one per column) Non-linearity defined based on the hysteretic rule with clear physical meaning Easy to control	Cannot be used for the analysis of bi-directional response Unable to estimate a local response.
Fiber element	Able to estimate the local response Can be used for the analysis of bi-directional response	Relatively complex analysis Several iterations are necessary to establish the appropriate model Control of results is more complex
MVL element	Relatively simple Non-linearity defined based on the hysteretic rule with clear physical meanings Able to estimate local response Can be used for analysis of bi-directional response	In general, several elements per column are necessary to obtain acceptable estimation of the response Appropriate number of elements should be defined iteratively

The presented models differ with respect to their sophistication. It can be concluded that although the beam-column element with lumped plasticity is the simplest, it is quite successful in the prediction of the global response. This makes it very suitable for the non-linear response history analysis, where the simple model is needed to simplify the analysis, less time-consuming, and easier to control. However, when the local (e.g. deformation or stress in some parts of the column's cross-section) or bi-directional response is investigated, this element cannot be used. In such cases the other two types of elements are preferred. The advantages and limitations of the three elements are summarized in the Table 2.2.

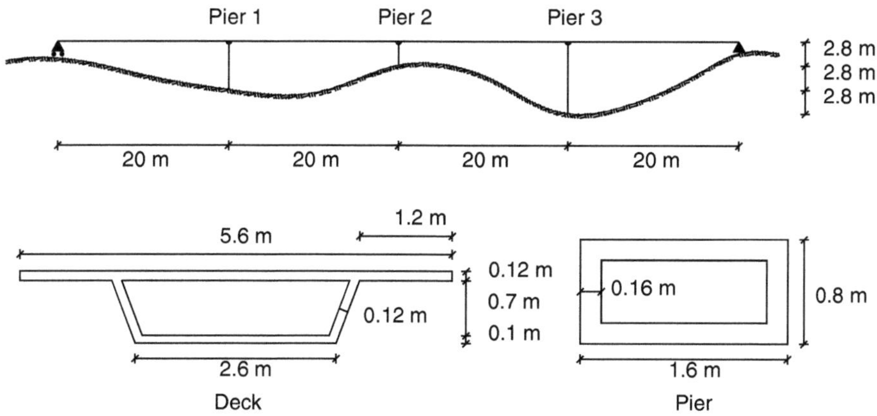

Fig. 2.36 Bridge configuration and member cross sections

2.5.6 Example 2 on Fiber Model Application

In the framework of an integrated European programme of research in support of Eurocode 8, six bridge prototypes, representative of typical multi-span continuous deck motorway bridges, have been designed (Pinto et al. 1996) with different procedures for a PGA of 0.35 g, in medium soil conditions (soil type B), applying the EC8 provisions. Corresponding large-scale (1:2.5) bridge models have then been constructed and tested in pseudo-dynamic (PsD) fashion at the Joint Research Centre at Ispra, Italy.

A PsD test, despite being carried out quasi-statically, employs on-line computer calculations and control together with experimental measurement of the properties of the real structure to provide a realistic simulation of its dynamic response. Inertial and viscous damping forces are modelled analytically, and an earthquake ground acceleration history is given as input data to the computer running the pseudo-dynamic algorithm. The horizontal displacements of the controlled degrees of freedom are calculated and then applied to the test structure by servo-controlled hydraulic actuators fixed to the reaction wall. The PsD testing of the bridge was performed using the sub-structuring technique, in which the piers were physically tested and the deck was numerically simulated online. Further details can be found in Pinto et al. (1996), Pinho (2000), and Sullivan et al. (2004).

2.5.6.1 The Bridge Model

One of the test models, labelled B213C, consists of three piers 5.6, 2.8 and 8.4 m high and a continuous deck with four identical 20 m spans. The deck is assumed to end at the abutments with shear-keys, but is free to rotate (Fig. 2.36). The deck-pier connections are assumed to be hinged transmitting lateral forces but no moments.

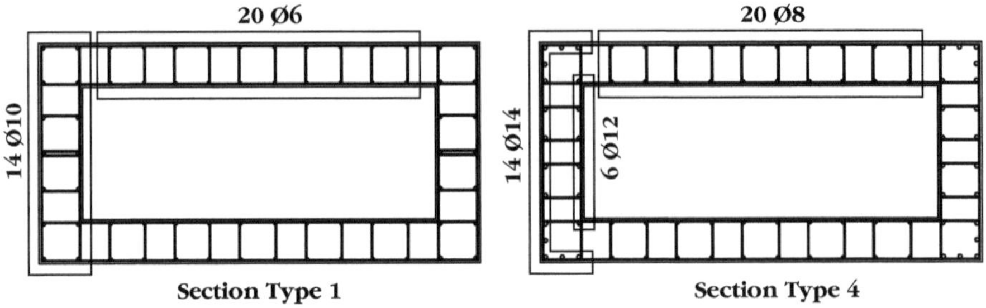

Fig. 2.37 Reinforcement layout

Table 2.3 Steel mechanical properties (Guedes 1997)

Diameter (mm)	Yield strength (MPa)	Ultimate strength (MPa)	Yield strain (%)	Ultimate strain (%)	Hardening
6	363.7	430.4	0.177	15.10	0.0022
8	503.4	563.0	0.244	12.30	0.0024
10	489.3	572.3	0.238	14.50	0.0028
12	558.2	646.8	0.271	12.80	0.0034
14	477.2	577.7	0.232	13.00	0.0038

Table 2.4 Summary of the pier cross section characteristics of the bridge (Guedes 1997)

Pier	Section type	Height (m)	Longitudinal steel (%)	Cubic concrete strength (MPa) (Compressive/tensile)
Pier 1	4	14	1.15	37.0/3.1
Pier 2	1	7	0.50	41.2/3.1
Pier 3	4	21	1.15	50.5/3.1

Table 2.5 Deck cross section geometrical and mechanical characteristics (Guedes 1997)

EA (kN)	EI_2 (kNm^2)	EI_3 (kNm^2)	GJ (kNm^2)
2.7837E + 07	1.3544E + 07	5.6517E + 07	2.8017E + 07

The piers have rectangular hollow section with 160 mm wall thickness (Fig. 2.36). The reinforcement layout of the pier models are shown in Fig. 2.37. The mechanical characteristics of materials (B500 Tempcore steel with E = 206 GPa for longitudinal rebars and C25/30 concrete) and the mechanical characteristics of the pier cross-sections are shown in Table 2.3, 2.4 and 2.5. The deck is a hollow-core prestressed concrete girder 5.6 m wide, as depicted in Fig. 2.36. In the PsD test, the deck was simulated numerically with 32 linear elastic Timoshenko eccentric beam elements, whose mechanical characteristics are presented in Table 2.5, where A is the cross-section area, I_2 and I_3 are the two moments of inertia with respect to the local principal axes, J is the torsional constant and E is the Young Modulus of 25 GPa. The inertia characteristics of the deck are based on a specific weight of 25 kN/m^3. The sub-structured part included a

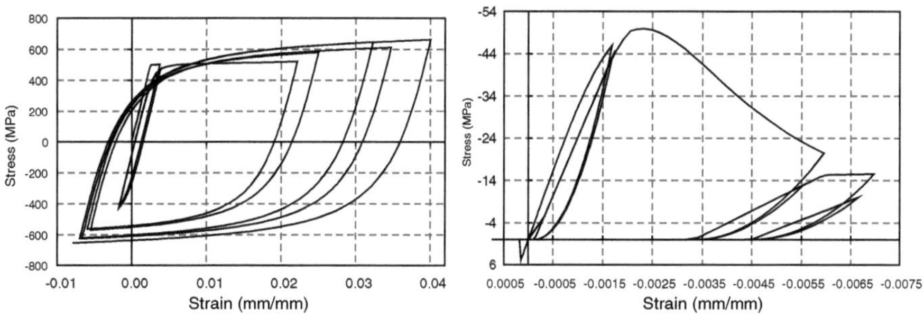

Fig. 2.38 Menegotto-Pinto steel model, with Filippou isotropic hardening (*left*), and nonlinear constant confinement concrete model (*right*)

Rayleigh damping matrix, featuring a damping ratio $\xi = 0.016$ associated to the two lower transversal natural frequencies of the complete bridge.

At the top of each pier, an axial force N = 1,700 kN was applied using actuators to simulate gravity load. The input ground motion was represented by a scaled accelerogram with duration of 4 s and a nominal peak acceleration of 0.875 g. Two pseudo-dynamic tests were performed on the structure: one with the input motion corresponding to the design earthquake and another defined on the basis of the estimated ultimate capacity of the bridges, and thus equal to 1.2 times the design earthquake.

2.5.6.2 Modelling the Piers in the FE Program

The computer program used in the analysis, SeismoStruct (SeismoSoft 2005), is a fibre–modelling FE package for seismic analysis of framed structures. The program is capable of estimating large displacement behaviour and the collapse load of frame structures under static or dynamic loading, accounting for geometric nonlinearities and material inelasticity. The piers have been modelled through a 3D inelastic beam-column element capable of capturing geometric and material nonlinearities. The number of fibres used in section equilibrium computations was 400.

The stress–strain behaviour of the steel (Fig. 2.38) has been described by the nonlinear model of Menegotto and Pinto (1973), as modified by Filippou et al. (1983) to include isotropic strain hardening. This is an accurate and convenient model due to its computational efficiency and its very good agreement with experimental results. It utilizes a damage modulus to represent more accurately the unloading stiffness, and has been modified and improved by Fragiadakis et al. (2008) to attain better stability and accuracy. The concrete has been represented through a nonlinear constant confinement concrete model (Fig. 2.38), as a good compromise between simplicity and accuracy: it is a uniaxial nonlinear model following the constitutive relationship proposed by Mander et al. (1988), later modified by Martinez-Rueda and Elnashai (1997) to improve numerical stability

Table 2.6 Parameters for the Menegotto-Pinto steel model, with Filippou isotropic hardening

Parameter	Sec 1	Sec 4
Modulus of elasticity (MPa)	203,000	203,000
Yield strength (MPa)	468	496
Strain hardening parameter	0.0027	0.0036
Transition curve initial shape parameter (default value)	20	20
1st transition curve shape coefficient (default value)	18.5	18.5
2nd transition curve shape coefficient (default value)	0.15	0.15
1st isotropic hardening coefficient (default value)	0.025	0.025
2nd isotropic hardening coefficient (default value)	2	2

Table 2.7 Parameters for the nonlinear constant confinement concrete model

Parameter	Pier 1	Pier 2	Pier 3
Cylinder compressive strength (MPa)	31.5	35.0	42.9
Tensile strength (MPa)	3.1	3.1	3.1
Strain at unconfined peak stress (m/m)	0.002	0.002	0.002
Constant confinement factor	1.2	1.2	1.2

under large deformations. The model calibrating parameters, fully describing the mechanical properties of steel and concrete, have been set as shown in Tables 2.6 and 2.7, where the concrete cylinder strength was estimated as being 85% of the cubic sample strength listed in Table 2.4.

2.5.6.3 Cyclic Response

After the PsD testing of the bridges, the tall and the medium piers were tested cyclically until failure (Pinto et al. 1996; Guedes 1997), respectively up to 230 and 150 mm of top displacement, under the imposed displacement history shown in Fig. 2.39. Additional cyclic tests (up to 72 mm at the top of the pier) were carried out on a short pier similar to the one tested in pseudo-dynamic fashion. These cyclic tests on the piers are numerically reproduced herein through a static response-history analysis, so as to enable a first check on the accuracy of the model. The numerical reproduction of the cyclic test has been performed imposing on the piers the displacement history resulting from the PsD test (Fig. 2.39). In addition, the steel young's modulus of the medium-height pier was halved, as suggested by Pinto et al. (1996), in order to reproduce the reduction in the stiffness due to the shear damage that this pier suffered prior to this cyclic test. No reduction in the steel properties was applied to the tall pier because it was not damaged during the PsD test.

Figure 2.40 shows a reasonable match between the experimental and numerical results for the medium pier. Only the reduction in strength at the very last cycle, when failure occurs, is not perfectly captured. Similar trends were observed in the response of the tall and short piers. It is clear that pinching has not been perfectly captured due to absence of shear deformation modelling in these fibre elements, an aspect still in the development stage (Ceresa et al. 2007). Had nonlinear shear

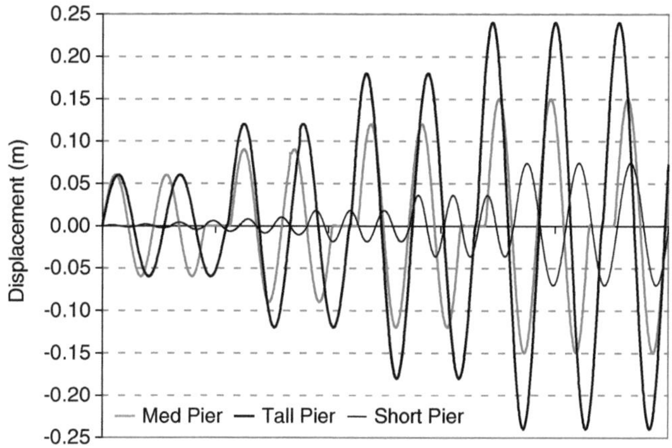

Fig. 2.39 Cyclic test displacement histories

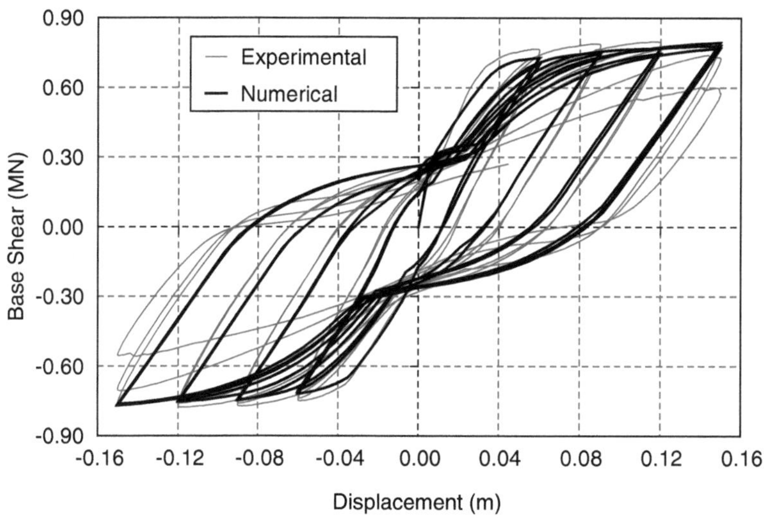

Fig. 2.40 Cyclic test results for the medium pier

deformations been considered, a closer match would have been achieved (e.g. Ceresa et al. 2009), but the computer analysis time would have been longer (Fig. 2.40).

2.5.6.4 Pseudo-dynamic Results

To analyze the pseudo-dynamic response of the piers, a complete FE model of the case-study bridge was created (Casarotti and Pinho 2006) and the calculated and measured results were compared in terms of displacements and forces at the top of the piers under the second, and stronger input earthquake motion.

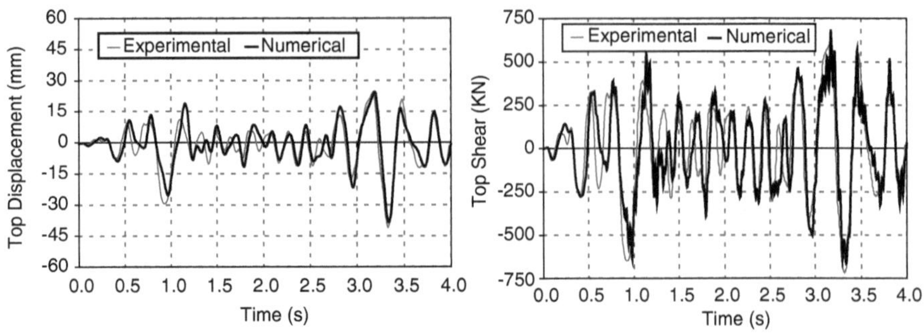

Fig. 2.41 Medium pier top displacements (*left*) and top shear (*right*)

Table 2.8 Ratios of the absolute maximum response obtained from numerical calculation to that from tests

	Tall pier (%)	Med pier (%)	Short pier (%)
Displacement	88	94	102
Top shear	90	95	188

Figure 2.41 shows the results for the medium pier. It can be seen that there is a good agreement in terms of both the amplitude and the frequency content of the response. The agreement is representative of the correlation for the other two piers. Table 2.8 lists the ratios of the maximum absolute response obtained from numerical calculation to that from the tests. It is noted that the force response of the squat pier is not reproduced with full accuracy, whereas displacements are very well predicted. The numerical overestimation of the action at the top of the short pier can be explained by the fact that the fibre-based element formulation did not account for shear deformation.

2.5.6.5 Closing Remarks

Structural behaviour is inherently nonlinear, particularly in the presence of large displacements or material nonlinearities, the structural response can be accurately caught only by means of nonlinear dynamic analyses. The fibre modelling approach employed in the current work is shown to be capable of associating simplicity of use, even for not highly experienced users. Moreover, its ability to simulate the nonlinear dynamic response of reinforced concrete bridges to seismic loads has been proven by simulating large-scale experimental pseudo-dynamic tests. Results of the dynamic and modal analyses performed reveal a good agreement with the pseudo-dynamic tests, both in terms of displacements and forces at the top of the tall and medium-height piers. At present, shear strains across the element cross-section are not included in the fibre-element formulation adopted, i.e. the strain state of a section is fully represented by the curvature at centroidal axial strains alone: this approach is not accurate enough for representing the squat pier deformation state, where shear deformations are of relevance. In this case, despite the relevance of the shear response, the prediction of the deformation of the squat member was still fairly good.

This section has thus illustrated how the use of simple-to-calibrate fibre structural models can be employed to reproduce with good level of accuracy the nonlinear structural response of continuous span bridge structures. In other words, it is believed that such an advanced analytical tool can be readily handled within a common engineering practice framework, provided a basic level of awareness on the decisions that the designer has to face, discussed herein is available.

2.5.7 Analytical Modelling of Hollow Box Columns

In Sects. 2.5.4–2.5.6 several modelling methods with different levels of sophistication were described and demonstrated. Another example of the development and application of microscopic models is presented in this section.

A series of quarter scale hollow columns were tested under cyclic loading at Porto University (Delgado et al. 2006, 2007, 2009). These columns were analyzed using CAST3M computer code (CEA 2003), which is a general purpose finite element analysis program. A wide variety of non-linear elements are included in CAST3M , particularly, a damage model developed at the Faculty of Engineering of Porto University (FEUP) (Faria et al. 1998; Costa et al. 2005). Studies have shown that the damage model is suitable for seismic behaviour analysis of RC bridge piers (Faria et al. 2004). The damage model is Continuum Damage Mechanics based constitutive model for the concrete zone discretized into 3D finite elements incorporating two independent scalar damage variables that account for the degradation due to tensile or compressive stress conditions. The Giuffré-Menegotto-Pinto model (Giuffrè and Pinto 1970) for the cyclic behaviour simulation of the steel reinforcement discretized via truss elements is used.

In analytical studies of the piers, the second and third repeated cycles of each displacement level were removed to facilitate comparison with experimental data. Only one-half of the cross section was modelled due to symmetry. In addition to comparing the measured and calculated hysteresis curves, the strains in various components were calculated and divided by the yield strain and are discussed in the following sections.

2.5.7.1 Comparison of Analytical and Experimental Results

Figures 2.42 and 2.43 compare the calculated and measured force-displacement relationships for two of the columns under moderate amplitude loading. It can be seen that generally good correlation was observed between the measured and calculated results.

The result of tensile damage pattern in PO1-N4 is illustrated in Fig. 2.43 for the initial cycles in which the first shear and flexural cracks were observed along nearly the entire pier height on the webs and concentrated at the flange bases. The compressive strain pattern is shown in Fig. 2.44b, c for 1.43% drift ratio when some damage was observed at the pier base during the tests. The deformed mesh for

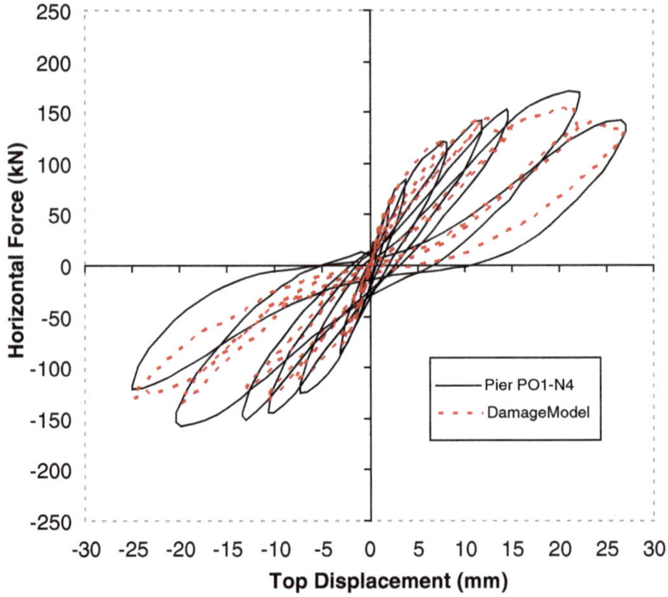

Fig. 2.42 Experimental and numerical results comparison for pier PO1-N4

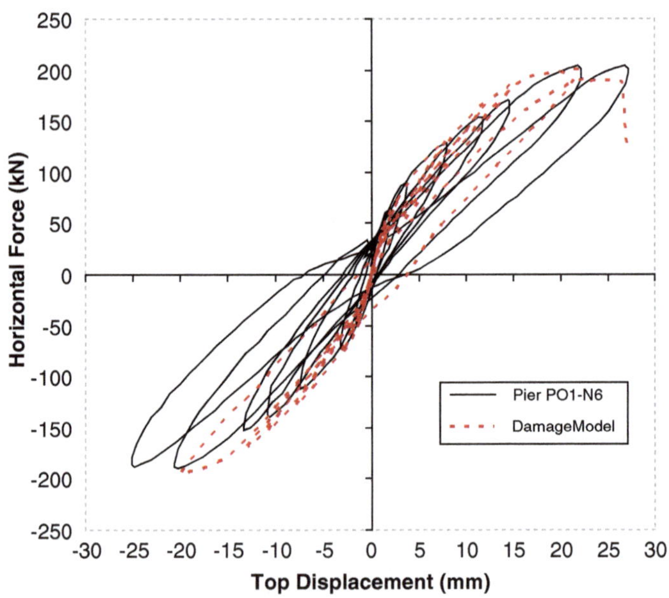

Fig. 2.43 Experimental and numerical results comparison for pier PO1-N6

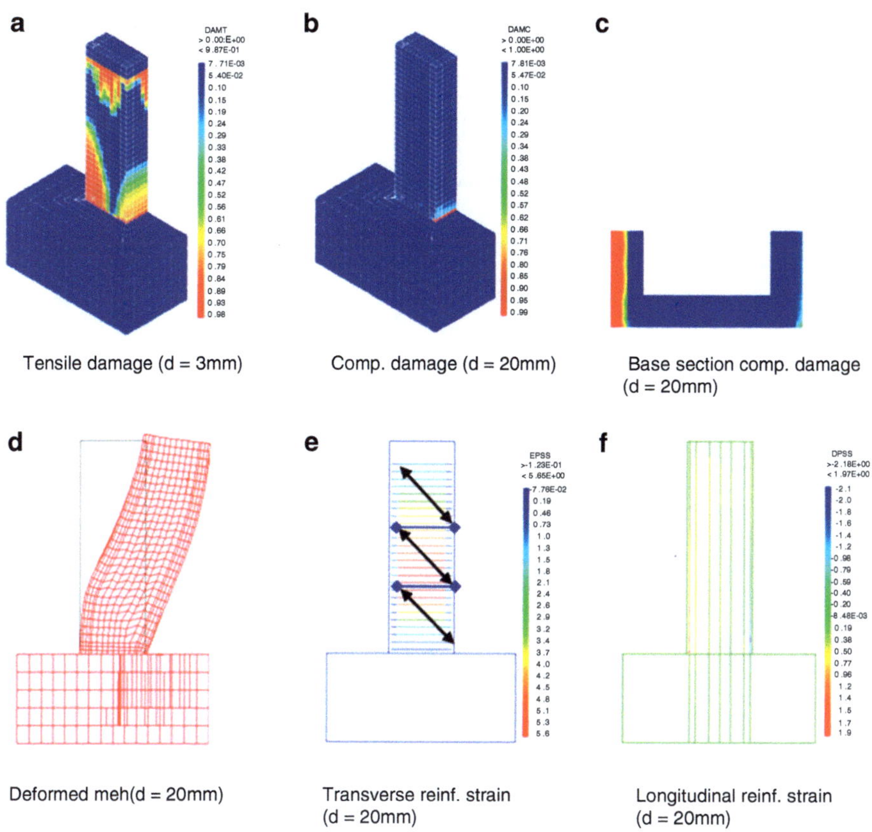

Fig. 2.44 Numerical results of PO1-N4 for 1.43% drift

the drift ratio of 1.43% is shown in Fig. 2.44d, in which significant shear deformation is evident. In the results shown in Fig. 2.44e, f, it can be observed that the strain in the transverse and longitudinal reinforcement bars under 1.43% drift ratio exceeded the yield strain as evidenced by ductilities higher than 1. The transverse reinforcement strain pattern shows the significant influence of the shear forces, reflected by stirrup yielding along the entire pier height suggesting development of a strut-and-tie shear mechanism. The longitudinal bar strains indicate yielding in the outer bars above the foundation, but with a maximum ductility of slightly above 1 suggesting very small flexural plastic deformations. These results are in general agreement with those observed in the experimental studies.

The calculated tensile strain ratios for the initial cycles (Fig. 2.45a) show significant cracking at webs and flanges, mainly concentrated at the pier base. The compressive damage pattern for 1.43% drift ratio (Fig. 2.45b and c) indicates maximum damage near the pier base, which is consistent with observed damage that consisted of minor concrete spalling at the corners. The deformed mesh for 1.43% drift ratio at the top of pier PO1-N6 shows significantly less shear deformation compared with PO1-N4 (Fig. 2.45d) because of the higher transverse steel ratio.

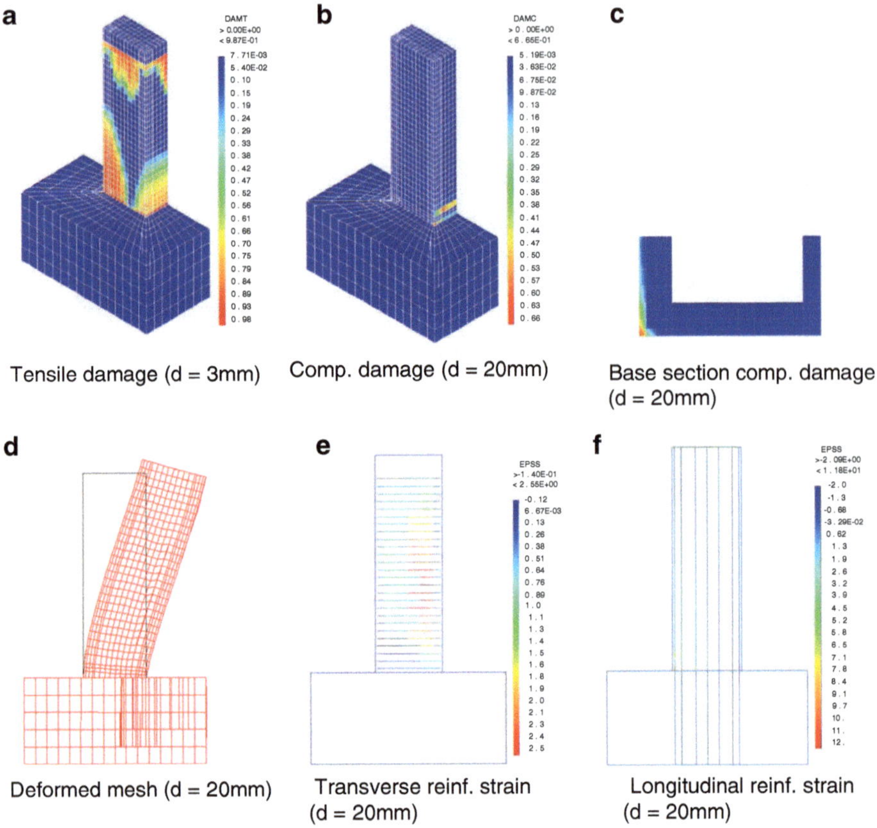

Fig. 2.45 Numerical results of PO1-N6 for 1.43% drift

Compared to transverse bar strains in PO1-N4, the strains shown in Fig. 2.45e are considerably lower due to the higher amount of transverse steel in PO1-N6. Finally, the longitudinal rebar strain distribution shows that extent of yielding is significantly higher than that of PO1-N4. This because the higher amount of transverse steel reduced shear degradation and allowed for higher flexural strains to develop.

2.5.7.2 Concluding Remarks

An application of refined constitutive models for estimating the nonlinear seismic behaviour of hollow box reinforced concrete bridge piers was described. A constitutive model based on the Continuum Damage Mechanics was used for the concrete, incorporating two independent scalar damage variables to reproduce the degradation under tensile or compressive stress conditions. Steel reinforcement was discretized via 2-noded truss elements, and the corresponding behaviour was simulated using the Menegotto-Pinto cyclic model.

The efficiency of the numerical model was demonstrated by simulating the experimental tests performed at the LESE on reduced scale bridge pier models tested under cyclic loads. Important characteristics of the test models were adequately captured using the analytical model. The shear capacity of each pier was also accurately captured by the model developed by Priestley et al. (1996), with shear failure in PO1-N4, and limited flexural ductility in PO1-N6, which has twice as much shear reinforcement. The ability of the detailed analytical model to capture shear deformations is due to the refined element modelling it incorporated. Distributed shear degradation and failure is generally hard to be replicated by simple macromodels.

2.6 Modelling of Dynamic Interaction Between Piers, Foundation and Soil

2.6.1 *Pseudo-static Winkler Approach*

The most commonly adopted engineering method for calculating the pseudo-static interaction between the piles of a bridge foundation and the soil is the Winkler model in which the soil reaction to pile movement is represented by independent (linear or non-linear) unidirectional translational spring elements distributed along the pile shaft to account for the soil response in the elastic and inelastic range respectively. Although approximate, Winkler formulations are widely used not only because their predictions are in good agreement with results from more rigorous solutions but also because the variation of soil properties along the pile length can be relatively easily incorporated. Moreover, they are efficient in terms of computational time required, thus allowing for easier numerical handling of the structural inelastic response.

The corresponding mechanical parameters for the springs are frequently obtained from experimental results (leading to P-y curves for lateral and T-z curves for axial loading) as well as from very simplified models. A commonly used P-y curve is the lateral soil resistance vs. deflection relationship proposed by the American Petroleum Institute (1993):

$$P = 0.9p_u \tanh\left[\frac{kH}{0.9p_u}y\right] \tag{2.19}$$

where p_u is the ultimate bearing capacity at depth H, y is the lateral deflection and k is the initial modulus of subgrade reaction which is both depth and diameter-dependent despite the fact that in many cases (i.e. NAVFAC 1982) the modulus of the subgrade reaction is assumed to be independent of diameter.

This curve is widely used in current practice, especially in the U.S., and it is applied also for dynamic or non-linear problems. The latter is achieved with a simple transformation of Eq. 2.1 to a bi-linear relationship by assuming a specific deformation Δ_y to enter the inelastic range (typically equal to 2.5 cm for

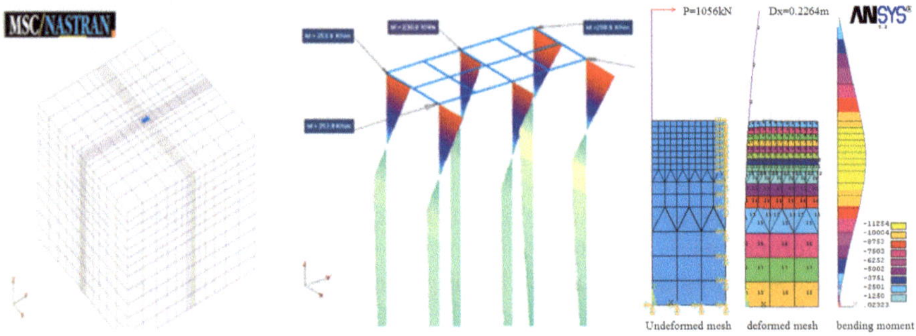

Fig. 2.46 Two and three dimensional finite element representation of soil-foundation-pier system stiffness (Kappos and Sextos 2001)

cohesionless soils) and a second branch stiffness reduced to 1/4 of the initial soil stiffness (Kappos and Sextos 2001). Alternatively to the above procedure, the static stiffness detached from the complex dynamic stiffness matrix (as discussed in the following section) is also used in practice.

To complete the foundation modelling, a horizontal inelastic soil spring can be used at the top of the pile to represent the strength and stiffness provided by passive soil resistance against the pile cap while a vertical inelastic spring is commonly used at the pile tip to account for downward and upward capacity of the supporting soil.

For the case that particular soil layers are considered to be susceptible to liquefaction, recent studies suggest that both the lateral subgrade reaction of piles and the maximum reaction force of the laterally-spreading soils have to be reduced at the corresponding locations along the pile length. This reduction factor lies in the range of 0.1–0.2 (Finn 2005) or 0.05–0.2 (Tokimatsu 1999). An average 10% therefore of the lateral stiffness provided along the liquefied soil layers is deemed reasonable, however, the designer has to bear in mind that the particular assumption is strongly earthquake magnitude dependent.

Based on the above discussion, it can be claimed that at least from a static point of view, the use of lateral soil resistance-deflection curves (even linear) is a convenient approach for the estimation of the dynamic characteristics of the bridge. Nevertheless, despite the wide application of the P-y approach for the assessment of the structural response in the design practice, there are certain limitations that have to be stressed:

(a) the uncertainty of estimating the parameters involved when load tests are not available (especially of defining p_u and k), is disproportionally high compared to the simplicity of the approach. It is notable that although Eq. 2.19 is adopted by both the Multidisciplinary Centre for Earthquake Engineering Research and ATC (MCEER/ATC 2003) and the California Department of Transportation (CALTRANS) guidelines, the proposed sets of the required subgrade moduli differ on average by a factor of 4 (Finn 2005).
(b) the relationships available that relate the 1D (expressed in terms of modulus of subgrade reaction k) to 2D and 3D soil stiffness (Fig. 2.46), the latter expressed

in terms of modulus of elasticity E_s, and Poisson's ratio v, are not directly related nor verified by 2D and 3D FE analyses. As a result, a set of calibration assumptions is required for establishing a correspondence between the Winkler and plane-strain FE approaches (Kappos and Sextos 2001) based on the initial formulations proposed by Vesic (1961):

$$k = Dk_h = \frac{0.65E_s}{(1 - v^2)^2} \sqrt[12]{\frac{E_s D^4}{(1 - v^2)E_p I_p}} \tag{2.20}$$

where k_h is the modulus of subgrade reaction and D the pile diameter. Moreover, the transformation from one soil parameter to the other is straightforward only in the case that they are assumed constant with the depth while a set of additional and rather case-dependent calibrations is required to obtain agreement in the inelastic range.

(c) the pile group effect is essentially neglected. Even if the piles are statically connected using appropriate single valued springs to represent the increased flexibility of a pile group compared to the summation of the stiffness of all individual piles (a practice that is acceptable for static analysis), the actual dynamic impedance and the subsequent damping are completely neglected.

(d) the extension of the above P-y relationships or other curves calibrated from static analysis or loading testing for use in the framework of dynamic analysis is subjective.

(e) estimating the effect of soil-structure interaction solely on the basis of the increased foundation flexibility, is an oversimplification that may lead to unconservative response estimates under certain circumstances (Mylonakis and Gazetas 2000).

(f) the convenience of the (particular statically based) P-y method, often leads to the extension of its application for the case of inelastic dynamic analysis in the time domain. Such an extension, although tempting for special cases of structural design (i.e. performance based design of new or retrofit of existing important structures) leads to the misleading perception of modelling refinement without proper understanding and consideration of the complex dynamic nature of SSI phenomena. As a result, important aspects of the soil-foundation-superstructure system response are hidden under allegedly 'all-purpose' 3D linear/nonlinear, static/dynamic stick models.

It can be claimed therefore that as soil-foundation-structure interaction is a multi-parametric and strongly frequency-dependent phenomenon, it inevitably has to be seen from a dynamic point of view, through a very careful selection of FE models, associated parameters and modelling assumptions. As a result, the pseudo-static Winkler approach is deemed appropriate only for cases wherein:

(a) what is of interest is the identification of the dynamic characteristics of the overall soil-foundation-superstructure system and not the actual seismic response of the system in the time domain.

(b) a preliminary (standard or modal) pushover analysis is performed in order to quickly assess the inelastic mechanisms that are expected to be developed in the bridge under earthquake excitation.
(c) a response spectrum analysis is performed or a linear response history analysis is run for a relatively low level of seismic forces.
(d) an inelastic dynamic analysis is conducted but the energy absorption is expected to be mainly concentrated on the superstructure while the material and radiation damping at the soil-foundation interface is a-priori judged of secondary importance (i.e. in cases that the underlying soil formations are stiff and uniform with depth).

2.6.2 Linear Soil-Foundation-Bridge Interaction Analysis in the Time Domain

Currently, despite the lack of specific design guidelines on how to model, compute and consider Soil-Structure Interaction effects in the framework of the seismic excitation of bridges in the time domain, it is common knowledge that foundation is flexible, dissipates energy and interacts with the surrounding soil and the superstructure, in such a way that it both filters seismic motion (kinematic interaction or wave scattering effect) while it is subjected to inertial forces generated by the vibration of the superstructure (inertial interaction). This phenomenon is complex and its beneficial or detrimental effect on the dynamic response of the structure is dependent on a series of parameters such as (Pender 1993; Wolf 1994) the intensity of ground motion, the dominant wavelengths, the angle of incidence of the seismic waves, the stromatography, the stiffness and damping of soil, as well as the size, geometry, stiffness, slenderness and dynamic characteristics of the foundation and the structure.

The basic methods to be used in the analysis of Soil-Structure Interaction effects that implement Finite Element Discretization are the Complete Finite Element approach and the Substructure Method. Due to the substantial computational time required, the Complete Finite Element approach is primarily used when the overall 2D or 3D geometry of the problem is of interest and the response of the soil-foundation-superstructure is mainly linear elastic. It has to be noted though, that in general, the use of Finite Element representation as a means of numerically predicting the seismic wave propagation, is relatively inferior to more specialized approaches such as the Finite Difference method. In order to enhance the accuracy of the FE analysis therefore, it is deemed critical to ensure realistic boundary conditions and optimal mesh dimensions. The first is commonly tackled with the incorporation of Kelvin elements (separately for the horizontal and vertical directions) which is easier to be implemented compared to other artificial boundary conditions such as superposition, paraxial or extrapolation boundaries. It is noted that with proper selection of model parameters, the boundary interference may indeed be eliminated, as has been shown by comparing numerical with experimental results (i.e. Pitilakis et al. 2004). The frequency dependent constants of the Kelvin elements can be calculated using the solution developed by Novak and Mitwally (1988), with coefficients proposed for the two horizontal and the vertical direction respectively. Additionally, it is important to achieve adequate (albeit not exhaustive) mesh refinement in order to minimize the error related to finite wave propagation. A very practical limit for the characteristic length l_c of the elements has been proposed by Lysmer and Kulemeyer (1969) and is still extensively used in FE discretization.

According to the Substructure Method on the other hand, the soil-foundation-structure system is divided into substructures, typically the superstructure, the near field soil domain inclusive of the foundation and the far field domain. Due to considerable computational economy the substructure method has been more extensively used in the past either in the form of:

(a) *Coupled FEM/BEM approaches* (Renault and Meskouris 2004; Savidis et al. 2000 among others): the advantage of this approach is that the soil can be discretizised only in the interaction horizon, while the boundary conditions are consistent and hence, the wave propagation in the free-field can be accurately calculated considering non-relaxed boundary conditions.
(b) *Uncoupling and superposition of kinematic and inertial interaction* (Kausel and Roesset 1994; Mylonakis et al. 1997): The dynamic stiffness matrix of the superstructure is attached to an additional impedance matrix representing the underlying unbounded soil region and the superstructure is then excited by the response history (denoted as Foundation Input Motion – F.I.M.) of a hypothetical soil-foundation sub-system lacking the superstructure mass.

In case of deep (pile) foundation, the procedure can be summarized in four independent steps:

1. Analysis of the free-field soil response (i.e. without the presence of piles) to vertically incident S waves.
2. Analysis of the interaction of the single pile or pile group with the surrounding soil, driven by the free-field response of step 1.
3. Computation of the dynamic impedances ("springs" and "dashpots") at the pile head or the pile-group cap, associated with the swaying (R_x and R_y), rocking R_{ry} and R_{rx}) and cross-swaying-rocking ($R_{x,ry}$ and $R_{y,rx}$) motion of the foundation. It is noted that specifically for pile groups, the dynamic impedance of the foundation cannot be computed by simply adding the dynamic stiffness of the individual piles. This pile-to-pile interaction is frequency-dependent, resulting from waves that are emitted from the periphery of each pile and propagate to strike the neighbouring piles (Mylonakis et al. 1997). A variety of numerical and analytical methods have been developed (Nogami et al. 1992; Makris and Gazetas 1992; El-Naggar and Novak 1996; Gazetas and Mylonakis 2002) to compute the dynamic response of this interaction. Especially for the rotational stiffness of foundations supporting bridge piers or building columns that are expected to develop plastic hinge at their base, a non-linear moment-rotation

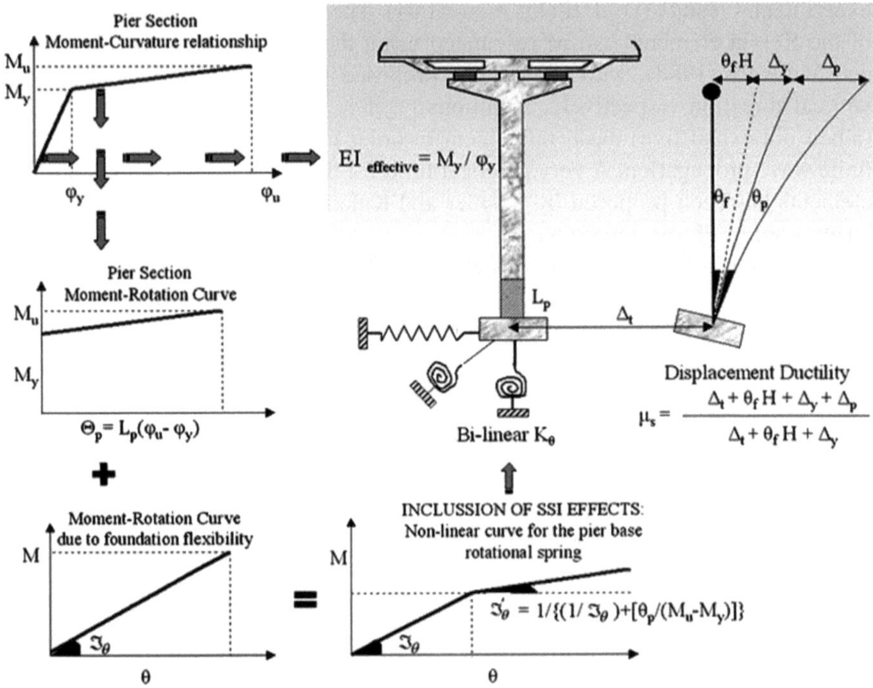

Fig. 2.47 Combination of soil-structure-interaction and post-yielding RC pier response

relationship can be implemented (Sextos et al. 2003a) combining the rotational compliance of the foundation with a lumped plasticity model for the RC section. In particular, by combining the flexibility of the non-linear pier-base inelastic spring and the linear rotational foundation spring that was calculated in the inertial soil-structure interaction stage, the final rotational spring (Fig. 2.47) is derived, being characterized by a first branch (uncoupled rotational) stiffness equal to$\Im_\Theta$ and a second branch stiffness $\Im'_\theta$equal to:

$$\Im'_\theta = \frac{1}{\frac{1}{\Im_\theta} + \frac{\theta_p}{M_u - M_y}} = \frac{1}{\mathrm{Re}\left[\frac{K_{HH}^{dyn} - K_{HM}^{dyn}/e}{K_{HH}^{dyn} \cdot K_{MM}^{dyn} - K_{HM}^{dyn\,2} + K_{\theta v}^{dyn} \cdot K_{HH}^{dyn}}\right] + \frac{(0.08L + 0.022 f_{yl} \cdot d_{bl}) \cdot (\phi_u - \phi_y)}{M_u - M_y}} \quad (2.21)$$

where θ_p, M_u, M_y are the plastic rotation, the ultimate and the yield moment of the pier base RC section respectively, $K_{HH}^{dyn}, K_{MM}^{dyn}, K_{HM}^{dyn}$ are the horizontal, rocking and coupled modes of vibration terms of the dynamic stiffness matrix, which for the case of pile groups are functions of the damping coefficients ζ_{HH}, ζ_{MH}, ζ_{MM} and the dynamic interaction factors α_{ij}^{dyn}, $K_{\theta v}^{dyn}$ is the (static) rotational stiffness component attributed to the antisymmetric vertical loading of the piles, $e = H/M$ is the foundation eccentricity, L is the distance from the critical pier section to the point of contraflexure, f_{yl} is the yield strength of the longitudinal bars and d_{bl} is the diameter of the longitudinal reinforcement.

4. Analysis of the dynamic response of the superstructure supported on the "springs" and "dashpots" of step 3, subjected to the kinematic pile-head motion of step 2.

The above approach is similarly applied for the case of surface foundations according to the corresponding four steps. A summary of closed-form solutions of dynamic stiffness of spread footings that have been derived by regression analysis, based on Finite- and Boundary-Element data can be found elsewhere (Mylonakis et al. 2006a). Commonly, due to their small embedment, the cross-swaying-rocking stiffness of the spread footings can be neglected.

In general, for each of the above analysis steps, several alternative formulations have been developed and published in the literature, including Finite Element, Boundary Element, semi-analytical and analytical solutions, as well as a variety of simplified methods (reviewed by Pender 1993 and Gazetas and Mylonakis 1998). In practice however, especially when specialized software is not available, it is convenient that the above inertial and kinematic interaction uncoupled sub-systems are split into two FE models using any standard structural analysis software and the corresponding damping and stiffness coefficients proposed in the literature.

2.6.2.1 Considering the Characteristics of Seismic Ground Motion

From the four steps described above, it is believed that the highest level of uncertainty is related to the identification of the incoming wavefield (as described in step 1). It has long been shown that the dynamic response of both sub-systems (i.e. massless soil-foundation and flexibly supported superstructure) is strongly frequency-dependent, hence earthquake- dependent. Therefore, as stressed by numerous previous studies, the study of SSI effects cannot be seen independently from the characteristics of the incoming seismic motion. To further illustrate the interplay between the supporting soil and foundation and the induced wavefield, Mylonakis et al. (2006b) have extended the examination of the role played by soil in the collapse of the Hanshin Expressway during the 1995 earthquake (Fig. 2.48). It is shown that the combination of the substantial modification of the bedrock motion frequency content by the underlying soil profile (apparent in Fig. 2.48) and the altered dynamic characteristics of the superstructure (both in terms of period elongation and enhanced fundamental mode contribution) lead to a 100% increase of the seismic demand compared to what would have been expected using a standard 'fixed base' approach.

Other studies (i.e. Sextos et al. 2003b) also verify that considering soil-structure interaction while neglecting local site response, the importance of SSI may be underestimated by a maximum factor of 2.5, compared to the case where the local (1D/2D or 3D basin site effects) had been properly incorporated in the framework of the SSI analysis.

As a result, it can be claimed that even the most state-of-the-art computing tools for considering the complex stiffness and damping matrix at the foundation level, when used independently of realistic consideration of the multi-layer, damped soil

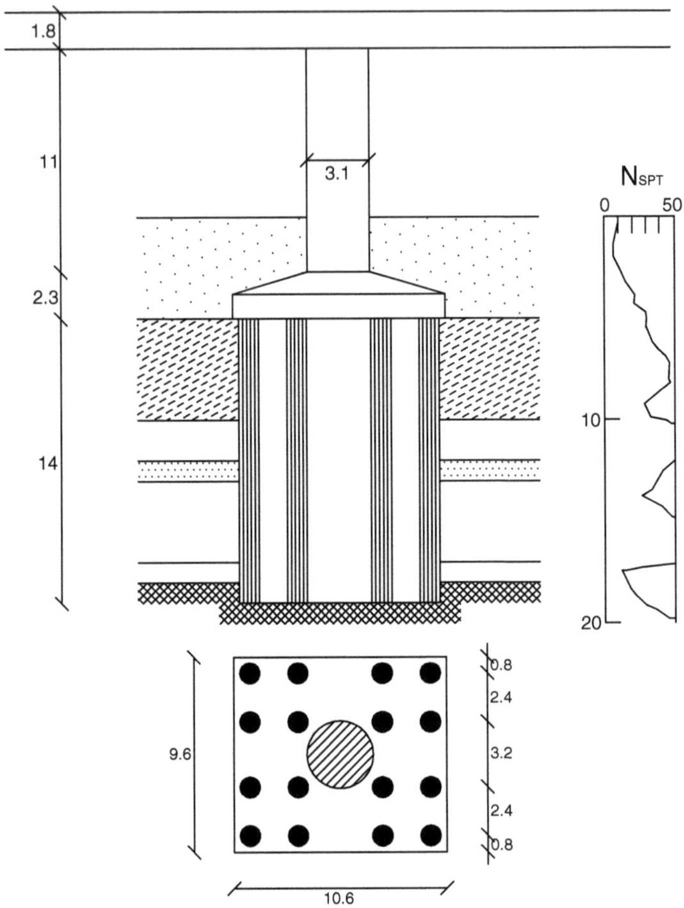

Fig. 2.48 Layout of the Hanshin expressway (Mylonakis et al. 2006b)

structure, may not only lead to erroneous results but most importantly, convey the false impression that advanced and refined modelling is used without ensuring the corresponding analysis reliability (Fig. 2.49).

2.6.2.2 Modelling Rotational, Directional and Multiple Support Excitation

Additionally to the importance of considering the local site response as part of the SSI analysis, there are at least three other significant factors that affect the excitation of the bridge:

(a) *Rotational component of the foundation response*: It has been shown (Kavadas and Gazetas 1993; Mylonakis et al. 1997) that the kinematic interaction between pile and soil, apart from filtering the high frequency content of the incident

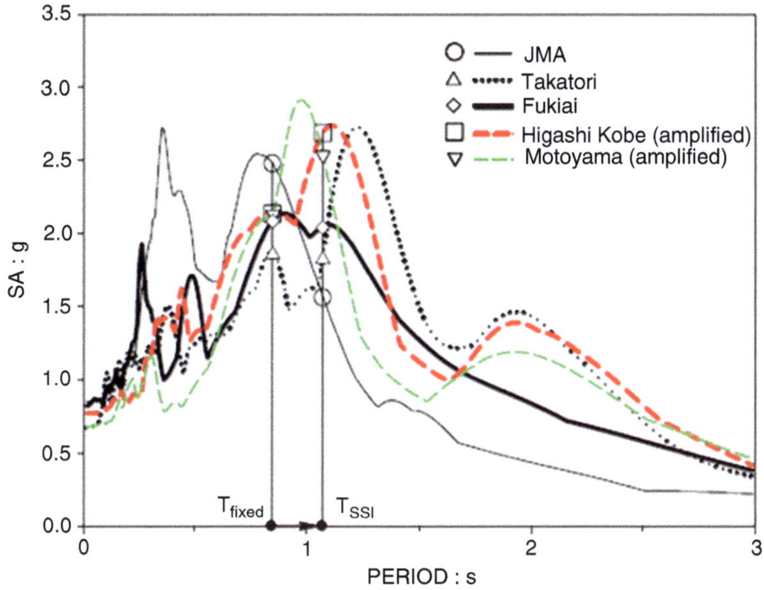

Fig. 2.49 Influence of period elongation on the attracted seismic force of the Hanshin expressway (Fukae section) for selected response spectra of the Kobe earthquake (Mylonakis et al. 2006b)

seismic waves, induces rotations at the pile head-base of the structure interface, especially in the case of single piles or footings with a very small number of piles (micropiles). This is a crucial issue, commonly neglected during the excitation of the superstructure, that has to be more thoroughly investigated.

(b) *Asynchronous excitation*: The spatial variation of earthquake ground motions is another aspect that is not accounted for in the vast majority of the design cases. This is of course a reasonable approximation for buildings or short bridges, but may considerably alter the overall seismic response of longer bridges especially in the case that the local site conditions vary significantly with length (Der Kiureghian, Keshishian 1997; Shinozuka et al. 2000; Mylonakis et al. 2001; Sextos et al. 2003b; Ates et al. 2006). This problem is very complex and various methods have been proposed for tackling the uncertainty related to the definition of a 'realistic' support-dependent earthquake input, as well as for predicting the conditions under which the effect of asynchronous input is indeed critical for the structure. Presenting in depth the multiparametric nature of spatial variability is beyond the scope of this book. However, it is interesting to notice that as, in most practical situations of long structures, the foundation soil and the subsequent foundation design varies significantly along the bridge length, the kinematic interaction is support-dependent, and hence the foundation presence is an additional source of spatial variability of ground motion. As a result, SSI effects and the final spatial variation of the travelling earthquake motion are strongly coupled phenomena.

(c) *Direction of excitation*: The importance of the angle of ground motion incidence has been highlighted by many researchers, working primarily on buildings and the Penzien and Watabe model (1975) that is, assuming the translational components of ground motion uncorrelated along a well-defined orthogonal axes of the structure. Recent research by Sextos et al. (2004; Sextos and Kappos 2008) among others, has also shown significant discrepancy between the response quantities produced when the seismic components are applied along the structural axes and those resulting from the excitation of the structure for different angles of wave incidence, especially in the case of long and curved bridges.

2.6.3 Nonlinear Soil-Foundation-Bridge Interaction Analysis in the Time Domain

2.6.3.1 Simulating the Non-linear Response of Soils

As the earthquake intensity increases, the behaviour of soil deposits becomes more and more complex, while the corresponding seismic demand to the bridge superstructure may increase substantially. Advanced FE software currently supports a large number of sophisticated material models most of which can be incorporated in the framework of 2D or 3D discretization. Nevertheless, the numerical simulation of the inelastic mechanisms developing simultaneously at the soil, foundation and structure is very demanding regarding the computational time required, whereas the accuracy foreseen through the use of advanced models is rather counteracted by the numerous uncertainties involved. This may be primarily attributed to the fact that:

(a) despite the significant advances in modern computer software, it is difficult to achieve a balance between the material modelling refinement for the soil and the structure as most computer codes do not support built-in features specialized for both soil volumes, and structural yielding mechanisms (RC, steel or masonry). Most importantly, even when they do (i.e. ANSYS, ABAQUS, ADINA among others), their simultaneous activation in the time domain may lead to significant analysis convergence difficulties.
(b) most material constitutive laws (i.e. elastic–perfectly plastic von Mises model, Mohr-Coulomb or the Drucker- Prager criterion) that are indeed implemented in the available structural or general purpose analysis software, require a number of decisions on particular coefficients to which quite often the analysis is found to be numerically sensitive. As a result, a large number of parametric analyses and verification studies are needed.
(c) even though general purpose numerical analysis software can now efficiently solve non-linear boundary conditions problems (i.e. sliding at the pile-soil interface or sliding/rocking at the footing soil interface), it is not by any means (and could not easily be) ensured that the coupled material and boundary conditions non-linearity is effectively predicted by the use of the two software features used simultaneously.

(d) similarly, the application of complex material laws for both pseudo-static and cyclic dynamic analysis does not lead to stable and accurate converged solutions unless first verified by very experienced users.

One solution to the above described problem is to assess in advance the locations of the soil-foundation-superstructure system that are expected to be subjected to the highest level of inelastic demand and develop the FE modelling strategy accordingly. As a result, it is deemed a reasonable engineering approximation for cases of low to moderate non-linearity in the soil, to apply the kinematic and inertial procedure described in the previous section (which, strictly speaking, is not valid in the inelastic range) using effective shear strain-compatible soil moduli and damping for the determination of the spring and dashpot coefficients (i.e. Makris and Gazetas 1992).

Whenever there is evidence of significant non-linear soil behaviour, and in order to avoid undesirable analysis complexity, it is proposed to combine refined 3D structural models developed primarily for the inelastic analysis of the superstructure in the time domain, with simple stick foundation elements supported on ready-to-use (and computationally non-demanding) nonlinear Winkler springs that account for the static, cyclic, and dynamic response of foundations.

Finally, when the response of the foundation itself is also sought, (i.e. pile plastic hinges are expected to develop below the surface), then the modelling refinement can extend to the foundation which can again be assumed supported on analytically derived springs that account for soil non-linearity.

Three dimensional inelastic analyses in the time domain for the complete soil-foundation-structure system is recommended mainly for cases that the structural response can be approximated as linear elastic and the soil plastic strains are of interest.

2.6.3.2 Dealing with Layer-Specific Liquefaction

One of the non-linear soil effects, which are most difficult to simulate using standard finite element software, is soil liquefaction. This is a crucial soil response phenomenon that may lead to large and permanent ground displacements hence it has to be properly considered at least for particular soil stratification. Nevertheless, with the exception of user-defined modules imported in commercial general purpose FE codes, the programs available for considering soil liquefaction do not account for structural inelastic behaviour (or do not account for the superstructure at all), with the exception of Opensees.

One possible approximation to overcome the particular problem is the implementation of an independent site response analysis where the liquefaction susceptible layers are appropriately accounted for through specialized software, thus leading to depth-dependent acceleration and displacement time histories (Kwon et al. 2009).

2.7 Modelling of Abutment-Embankment-Superstructure Interaction

2.7.1 *Simple P-y Relationships for Modelling Embankment-Abutment Systems*

Nonlinear static (pushover) analysis has become a popular tool for seismic assessment of reinforced concrete (RC) bridges primarily as a means to identify the hierarchy of failure at a low computational cost. Nevertheless, the nonlinearity expected in bridges during strong ground motions, cannot be attributed solely to yielding of RC sections, although these are the elements that are often purposely designed to exhibit inelastic behaviour. However, additional material nonlinearity mechanisms (of the foundation, approach embankment, and/or backfill soil) and geometrical nonlinearity mechanisms (activation of control components such as bearings, 'stoppers', or seismic joints) can also contribute significantly to the nonlinearity of the overall system response. Typically, both sources of nonlinearity affect the seismic response of a bridge; however, hysteretic response of nonlinear materials may have higher levels of uncertainty compared to the (pre-defined) presence of gaps and joints. Despite the importance of modelling such complex bridge lateral boundary conditions and the existence of specific guidelines in the US (Caltrans, ATC, MCEER) and in Europe (Eurocode 8–2) for the design of pile foundations and abutments, only minor guidance is provided by codes for numerical modelling of the coupled soil-bridge system. Practical consideration and static assessment of the nonlinear soil-foundation-pier-deck system may be found in Elnashai and McClure (1996) and Kappos and Sextos (2001), among others, while static spring-based approaches for the study of soil-abutment-deck interaction have been long developed (Maragakis et al. 1989; Siddharthan et al. 1997; Spyrakos and Ioannidis 2003; Mackie and Stojadinovic 2002; Shamsabadi et al. 2007) most of which are summarized in a comprehensive report by Aviram et al. (2008). When the problem is studied under dynamic loading (Goel and Chopra 1997; Zhang and Makris 2001 and 2002; Kotsoglou and Pantazopoulou 2007 and 2009, among others) the derivation of the complex and coupled dynamic stiffness matrix of the lateral supports of the bridge is a multi-parametric and complicated process and as such, wider scatter is observed among the solutions available in the literature.

In addition, consideration of the 3D geometric and material nonlinearity of abutment-foundation-backfill-embankment system is often equally complex, uncertain, and computationally expensive and impractical for standard bridges. As a result, engineers resort to the simplified relationships prescribed by Caltrans to estimate the abutment-backfill soil capacity and stiffness. Another simplified alternative is carrying out a separate pushover analysis that can be performed for the abutment and foundation systems in order to quantify the lateral support stiffness of the bridge (Kappos et al. 2007). The advantage of this approach is that both soil and (abutment and foundation) concrete nonlinear behaviour can be considered as a

Fig. 2.50 Configuration of a typical seat type abutment-embankment system studied

means to provide case-specific force–deflection (i.e. P-y) relationships that can, in turn, be used as nonlinear spring boundary constitutive models in the pushover analysis of the overall bridge structure.

Based on the above observations, an effort was recently made (Sextos et al. 2008) to extend the above concept by performing a set of 3Dimensional nonlinear finite element analyses on typical California overpass abutment-embankment systems in order to provide simplified P-y relationships of the lateral supporting system as a function of abutment type, foundation-embankment-backfill geometry, and soil properties. These relationships can be potentially used in cases where more accurate data is not available.

2.7.2 Typical Bridges Studied

Six typical reinforced concrete California bridges (namely Route 14, LADWP, W180, MGR, Adobe, and La Veta) consisting of box-girder superstructures, seat-type abutments, and shallow pile foundations were adopted in the study by (Sextos et al. 2008). Given the short spans and relatively high deck stiffness of the particular structures, the embankment mobilization and the inelastic behaviour of the soil material under high shear deformation levels is anticipated to have a significant effect on the response of the bridge under seismic loading. The geometry of the cases studied is illustrated in Fig. 2.50 and is also summarized in Table 2.9.

2.7.3 Modelling of the Abutment-Foundation-Backfill-Embankment Systems

The abutment-embankment systems were modelled in 3Dimensional space using the finite element program ABAQUS and are presented in Fig. 2.51. It was deemed a realistic assumption to consider the soil (backfill, embankment and foundation) as

Table 2.9 Geometrical characteristics of the six abutment types

	H (m)	*h_1 (m)*	*h_2 (m)*	*b (m)*	*l (m)*	*L (m)*	*L_{ww} (m)*	*L_{pile} (m)*	*D_{pile}(m)*	*pile group*
Route 14	4.70	1.75	2.25	1.00	3.40	16.50	5.35	–	–	–
LADWP	4.00	1.30	2.10	1.00	2.90	13.50	4.80	–	–	–
W180	5.30	2.35	2.25	1.20	3.50	13.00	5.70	13.00	0.40	3 × 10
MGR	5.00	1.90	2.35	1.10	3.50	14.00	5.30	15.00	0.75	5 × 1/6 × 1
Adobe	4.40	1.25	2.25	1.20	2.45	13.00	6.00	15.00	0.40	2 × 10
La Veta	4.15	1.90	1.60	1.15	3.00	23.50	5.25	17.00	0.60	2 × 11

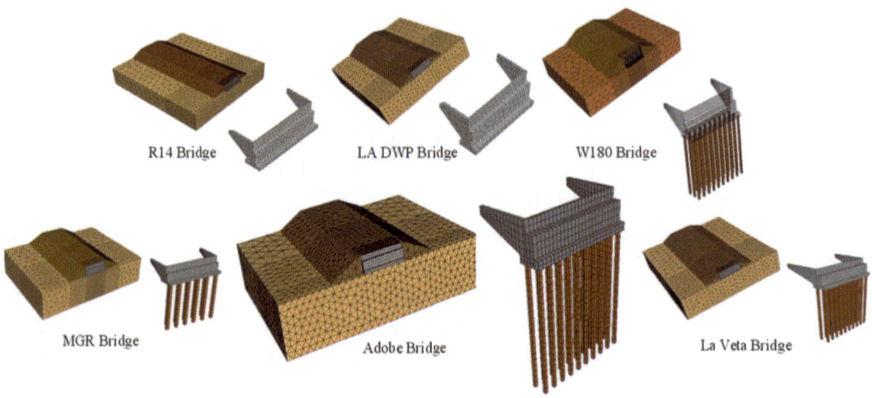

Fig. 2.51 Finite element models of the six abutment-embankment systems (Sextos et al. 2008)

the nonlinear material mechanism while the RC (abutment and pile) sections remain linear elastic with cracked section properties reduced by a factor of 2/3 compared to their gross section stiffness. Backfill soil was considered to be well compacted granular material according to the Caltrans guidelines despite the fact that in practice different soil types may exist. For modelling purposes, a Young's modulus of 60 MPa and a friction angle of 39° were adopted for the backfill soil type while, for simplicity, the same properties were assumed for the embankment as well, even though backfill material is usually stiffer and stronger than the embankment material. The foundation soil was parametrically modified to correspond to competent soil (A) and poor soil (B) according to Caltrans SDC (2006). Characteristic values of modulus of elasticity $E = 15$ MPa, friction angle $\varphi = 22^{\circ}$ and cohesion $c = 100$kPa for Soil A and $E = 5$ MPa, $\varphi = 5^{\circ}$, cohesion $c = 50$kPa for Soil B were also assumed for this study while Poisson's ratio (ν) and the unit weight (γ) were taken equal to 0.3 and 20kN/m^3 respectively. It is noted herein, that in the framework of the static analysis, compatible (Das 1994) static soil stiffness was adopted since the implementation of shear wave velocity typically overestimates the static stiffness of the finite element volume. The Mohr-Coulomb constitutive model implemented in ABAQUS was used to simulate the nonlinear soil behaviour while a gradually increasing pressure was applied for the push-over analysis as distributed normal and shear force on the abutment, along the longitudinal and transverse directions, respectively.

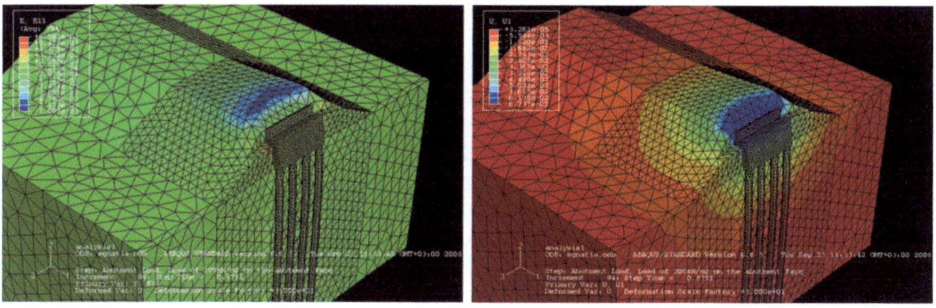

Fig. 2.52 Indicative plastic strains (*left*) and displacements (*right*) of an Egnatia Highway overcrossing (Sextos & Taskari, 2009)

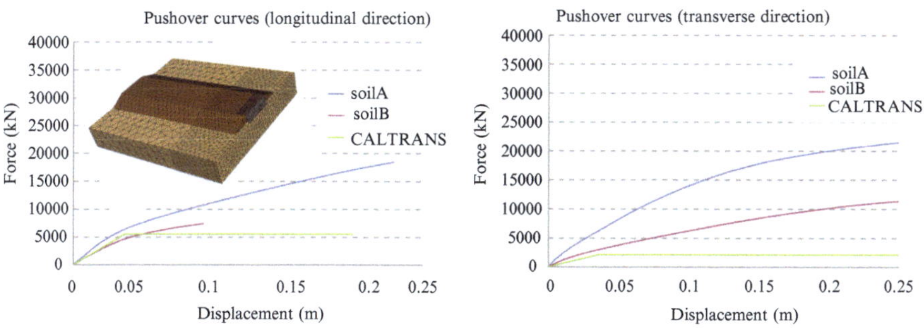

Fig. 2.53 Pushover curves of the Route 14 bridge abutment-embankment system along the longitudinal (*left*) and transverse direction (*right*)

Details regarding finite element discretization can be found in Sextos et al. (2008). The corresponding nonlinear P-y relationships were derived for the six bridges studied and for both soil types and excitation directions. It is noted herein that the particular approach is not affected by the potential presence of bearings, gaps, stoppers, and joints that have to be modelled independently as part of the main bridge structure, but rather represent solely the abutment-embankment system (Fig. 2.52).

2.7.4 Proposed P-y Relationships for Typical Abutment-Embankment Systems and Comparison with Caltrans Guidelines

Figures 2.53, 2.54, 2.55, 2.56, 2.57, 2.58 (Sextos et al. 2008) illustrate the pushover curves for the longitudinal (left) and the transverse (right) direction of the six abutment-embankment systems studied for the two Caltrans soil types (A and B). Based on passive earth pressure tests and the force deflection results from

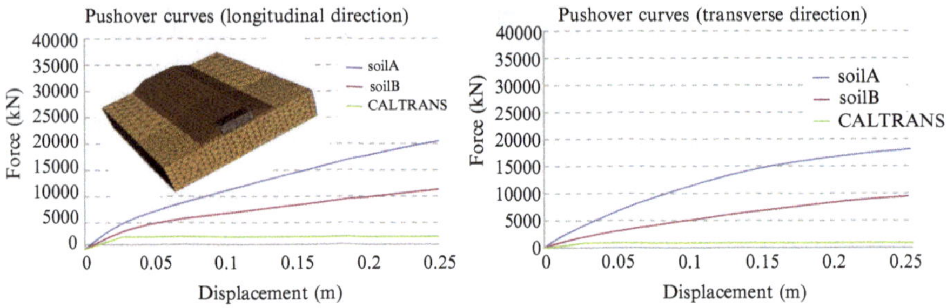

Fig. 2.54 Pushover curves of the LADWP bridge abutment-embankment system along the longitudinal (*left*) and transverse direction (*right*)

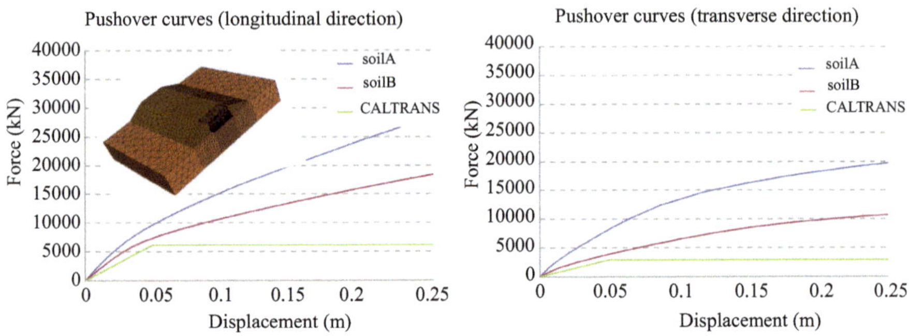

Fig. 2.55 Pushover curves of the W180 bridge abutment-embankment system along the longitudinal (*left*) and transverse direction (*right*)

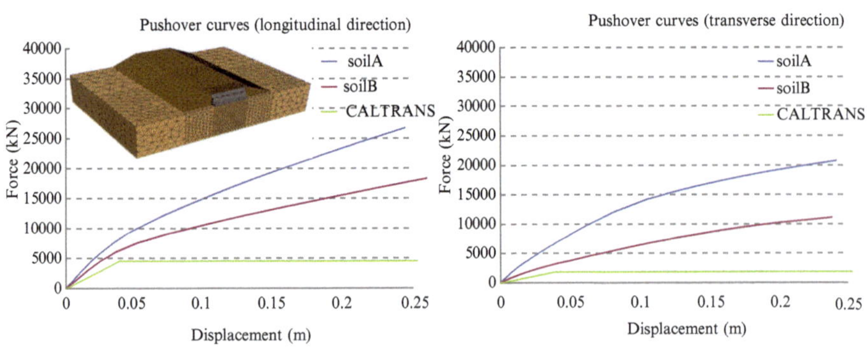

Fig. 2.56 Pushover curves of the MGR bridge abutment-embankment system along the longitudinal (*left*) and transverse direction (*right*)

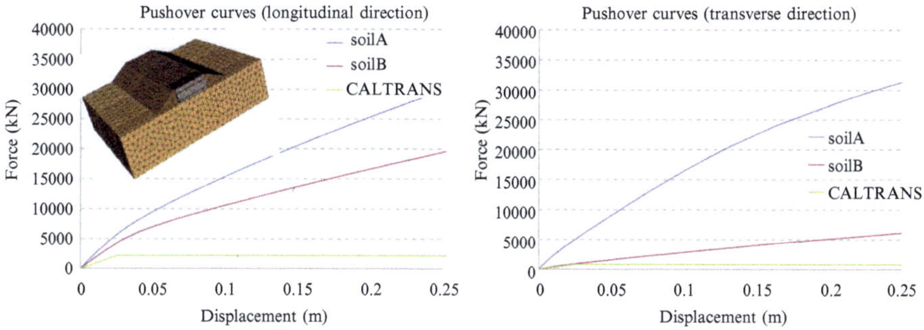

Fig. 2.57 Pushover curves of the Adobe bridge abutment-embankment system along the longitudinal (*left*) and transverse direction (*right*)

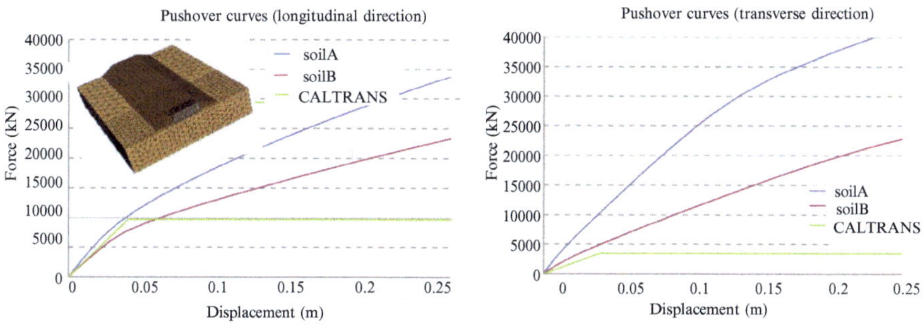

Fig. 2.58 Pushover curves of the La Veta bridge abutment-embankment system along the longitudinal (*left*) and transverse direction (*right*)

large-scale abutment testing at UC Davis, a value for the initial embankment fill stiffness equal to 11.5kN/mm/m width of the wall was used, as recommended by the Caltrans guidelines. The initial stiffness was adjusted proportionally to the backwall height:

$$K_{abut} = K_i \times w \times (h/1.7) \tag{2.22}$$

where *w* is the width of the backwall and (*h*/1.7) is the proportionality factor based on the 1.7 m height of UC Davis abutment specimen. The ultimate abutment load was assumed to be limited by a maximum static soil passive pressure of 239kPa. In the transverse direction, the abutment stiffness and strength obtained for the longitudinal direction were modified using factors corresponding to a wing wall effectiveness and participation coefficients of 2/3 and 4/3, respectively (Maroney and Chai 1994).

The initial stiffness and the ultimate capacity of the abutments according to Caltrans method (2006) are also plotted for comparison on each of Figs. 2.53, 2.54, 2.55, 2.56, 2.57, 2.58. The elastic and inelastic stiffness values for all the systems as

Table 2.10 Bilinear idealization of the pushover curves derived for the longitudinal (*left*) and the transverse direction (*right*) as normalized by the abutment width

LONG direction (MN/m^2)	Soil A $K_{initial}$ K_{inel}	Soil A $\delta_{yA}(m)$	Soil B $K_{initial}$ K_{inel}	Soil B $\delta_{yB}(m)$	Soil A $K_{initial}$ K_{inel}	Soil A $\delta_{yA}(m)$	Soil B $K_{initial}$ K_{inel}	Soil B $\delta_{yB}(m)$	TRANS direction (MN/m^2)
Route 14	15.7	0.037	11.2	0.033	12.3	0.037	5.4	0.033	*Route 14*
	4.4		3.4		2.8		1.9		
LADWP	18.2	0.041	12.6	0.038	12.9	0.041	6.1	0.038	*LADWP*
	6.1		3.0		3.3		2.6		
W180	28.7	0.05	22.8	0.046	21.1	0.05	9.3	0.046	*W180*
	10.8		6.3		4.7		2.6		
MGR	24.6	0.052	19.9	0.043	18.0	0.052	8.0	0.043	*MGR*
	8.4		5.3		4.6		2.9		
Adobe	21.8	0.047	16.5	0.043	17.1	0.047	3.2	0.043	*Adobe*
	9.6		5.9		8.1		2.1		
La Veta	14.8	0.048	11.5	0.042	12.7	0.048	5.7	0.042	*La Veta*
	5.1		3.5		4.7		3.4		

well as the corresponding yield displacements δy are summarized in Table 2.10, normalized by the width of the abutment so the resulting unit is [force]/[length]^2. As anticipated, the stiffness values (both initial and post-elastic) in the longitudinal direction are in general larger than those in the transverse direction. Moreover, the initial stiffness values obtained from 3D analysis are generally higher than those derived using the Caltrans procedure. The reason should be the fact that the latter does not account for factors such as abutment dimensions, embankment geometry, soil properties and foundation system stiffness but is based on a single, well-controlled, large-scale abutment test result. In addition, the 3D finite element approach is equally related to a number of modelling uncertainties.

References

Abdel-Mohti A, Pekcan G (2008) Seismic response of skewed RC box-girder bridges. J Earthq Eng Eng Vib, IEM and MCEER 7(4):415–426

ACI [American Concrete Institute] (2008) ACI 318–08 Building code requirements for structural concrete. American Concrete Institute, Farmington Hills

Al-Hussaini TM, Zayas VA, Constantinou MC (1994) Seismic isolation of a multi-story frame structure using spherical sliding isolation systems. Technical Report No. NCEER-94-0007, National Centre for Earthquake Engineering Research, Buffalo

Arêde A (1997) Seismic assessment of reinforced concrete frame structures with a new flexibility based element, Ph.D. thesis, FEUP, Porto. http://ncrep.fe.up.pt/web/artigos/AArede_PhD_Thesis_Public.pdf

Arêde A, Pinto AV (1996) Reinforced concrete global section modelling: definition of skeleton curves. Special Publication No.I.96.36, ISIS, EC, JRC, Ispra (VA), Italy

Arêde A, Vila Pouca N, Monteiro A, Delgado P, Costa A, Delgado R (2009) RC hollow-pier modelling and shear influence on the cyclic numerical response. In: Proceedings of the

COMPDYN 2009 – ECCOMAS thematic conference on computational methods in structural dynamics and earthquake engineering, Rhodes
Ates S, Bayraktar A, Dumanoglu A (2006) The effect of spatially varying earthquake ground motions on the stochastic response of bridges isolated with friction pendulum systems. Soil Dyn Earthq Eng 26:31–44
Aviram A, Mackie K, Stojadinovic B (2008) Guidelines for nonlinear analysis of bridge structures in California. PEER Report 2008/03, University of California, Berkeley
Biskinis D, Fardis MN (2007) Effect of lap splices on flexural resistance and cyclic deformation capacity of RC members. Beton- und Stahlbetonbau (Ernst & Sohn, Berlin), 102, 51–59
Bozorgzadeh A, Megally S, Restrepo J, Ashford SA (2006) Capacity evaluation of exterior sacrificial shear keys of bridge abutments. J Bridge Eng 11(5):555–565
Buckle IG, Constantinou M, Dicleli M, Ghasemi H (2006) Seismic isolation of highway bridges. Special Publication MCEER-06-SP07, Multidisciplinary Centre for Extreme Events Research, Buffalo
Caltrans [California Department of Transportation] (2006) Caltrans seismic design criteria version 1.4. Engineering Service Centre, Earthquake Engineering Branch, Sacramento, CA.
Casarotti C, Pinho R (2006) Seismic response of continuous span bridges through fibre-based finite element analysis. J Earthq Eng Eng Vib 5(1):119–131
CEA (2003) Manuel d'utilisation de Cast3m, Commissariat à l'Énergie Atomique, Pasquet P
CEB (1996) Costa and Costa model – RC frames under earthquake loading, Comité Euro-International du Béton, Bulletin n°231
Ceresa P, Petrini L, Pinho R (2007) Flexure-shear fibre beam-column elements for modelling frame structures under seismic loading – state of the art. J Earthq Eng 11(SP1):46–88
Ceresa P, Petrini L, Pinho R, Sousa R (2009) A fibre flexure-shear model for seismic analysis of RC framed structures. Earthq Eng Struct Dyn 38(5):565–586
Computers and Structures, Inc (2007) CSI analysis reference manual for SAP2000, ETABS, and SAFE. Computers and Structures, Inc, Berkeley
Constantinou, MC, Symans, MD, Tsopelas, P ,Taylor, DP (1993) Fluid viscous dampers in applications of seismic energy dissipation and seismic isolation. In: Proceedings of ATC-17-1 seminar on seismic isolation, passive energy dissipation, and active control, San Francisco, CA, pp 581–591
Costa AG, Costa AC (1987) Modelo Histerético das Forças-Deslocamentos Adequado à Anàlise Sísmica de Estruturas (Force-displacement hysteretic model for seismic analysis of structures). LNEC, Lisbon
Costa C, Pegon P, Arêde A, Castro J (2005) Implementation of the damage model in tension and compression with plasticity in Cast3m Report EUR, ISPC, CEC, JRC, Ispra (VA)
Das BM (1994) Principles of geotechnical engineering, 3rd edn. PWS Kent Publishers, Boston
Delgado P, Costa A, Delgado R (2002) A simple methodology for seismic safety assessment of bridges. In: 12th European conference on earthquake engineering, Elsevier Science Ltd, London, 9–13 September
Delgado P, Rocha P, Rodrigues V, Santos M, Arêde A, Vila Pouca N, Costa A, Delgado R (2006) Experimental cyclic tests and retrofit of RC hollow piers. In: Proceedings of the 13th European conference on earthquake engineering (13ECEE), 3–8 September, Paper N. 1205, Geneva, Switzerland
Delgado P, Rocha P, Pedrosa J, Arêde A, Vila Pouca N, Santos M, Costa A, Delgado R (2007) Retrofitting of bridge hollow piers with CFRP. In: COMPDYN 2007 – ECCOMAS thematic conference on computational methods in structural dynamics and earthquake engineering, 13–16 June, Paper N. 1492, Rethymnon, Crete, Greece,
Delgado R, Delgado P, Vila Pouca N, Arêde A, Rocha P, Costa A (2009) Shear effects on hollow section piers under seismic actions: experimental and numerical analysis. Bull Earthq Eng 7(2):377–389
Der Kiureghian A, Keshishian PE (1997) Effects of incoherence, wave passage and spatially varying site conditions on bridge response. In: Proceedings of FHWA/NCEER workshop on the national representation of seismic motion, Report 97–0010, New York, pp 393–407

Derham CJ, Kelly JM, Thomas AG (1985) Nonlinear natural rubber bearings for seismic isolation. Nucl Eng Des 84(3):417–428

Dolce M, Cardone D, Marnetto R (2000) Implementation and testing of passive control devices based on shape memory alloys. Earth Eng Struct Dyn 29:945–958

Dolce M, Cardone D, Croatto F (2005) Frictional behaviour of Steel-PTFE interfaces for seismic isolation. Bull Earthq Eng 3(1):75–99

Dolce M, Cardone D, Marnetto R, Nigro D, Palermo G (2003) A new added damping rubber isolator (ADRI): experimental tests and numerical simulations, Proc. 8th World Seminar on Seismic Isolation, Energy Dissipation and Active Vibration Control of Structures, Yerevan (Armenia), Oct 6–10

Duarte RT, Oliveira CS, Costa AC, Costa AG (1990) A non-linear model for seismic analysis, design and safety assessment of reinforced concrete buildings. In: Koridze A (ed) Earthquake damage evaluation & vulnerability analysis of building structures. OMEGA Scientific, Wallingford

El-Naggar MH, Novak M (1996) Nonlinear analysis for dynamic lateral pile response. J Soil Dyn Earthq Eng 15(4):233–244

Elnashai AS, McClure DC (1996) Effect of modelling assumptions and input motion characteristics on seismic design parameters of RC ridge piers. Earthq Eng Struct Dyn 25(5):435–463

Fajfar P, Fischinger M (1987) Non-linear seismic analysis of RC buildings: implications of a case study. Eur Earthq Eng 1(1):31–43

Faria R, Oliver J, Cervera M (1998) A strain based plastic viscous damage model for massive concrete structures. Int J Solid Struct 35(14):1533–1558

Faria R, Vila Pouca N, Delgado R (2002) Seismic behaviour of a RC wall: numerical simulation and experimental validation. J Earthq Eng 6(4):473–498

Faria R, Vila Pouca N, Delgado R (2004) Simulation of the cyclic behaviour of RC rectangular hollow section bridge piers via a detailed numerical model. J Earthq Eng 8(5):725–748

Filippou FC, Popov EP, Bertero VV (1983) Modelling of RC joints under cyclic excitations. J Struct Eng 109(11):2666–2684

Finn L (2005) A study of piles during earthquakes: issues of design and analysis. Bull Earthq Eng 3:141–234

Fischinger M, Vidic T, Fajfar P (1992) Non-linear seismic analysis of structural walls using the multiple-vertical-line-element model, non-linear seismic analysis and design of reinforced concrete buildings. Elsevier, Bled, pp 191–202

Fischinger M, Isaković T, Kante P (2004) Implementation of a macro model to predict seismic response of RC structural walls. Comput Concrete 1(2):211–226

Fragiadakis M, Pinho R, Antoniou S (2008) Modelling inelastic buckling of reinforcing bars under earthquake loading. In: Papadrakakis M, Charmpis DC, Lagaros ND, Tsompanakis Y (eds) Progress in computational dynamics and earthquake engineering. A.A. Balkema Publishers – Taylor, Francis, Rotterdam

Garstka B, Kratzig WB, Stangenberg F (1993) Damage assessment in cyclically reinforced concrete members. In: Proceedings of the second European conference on structural dynamics – EURODYN'93, 21–23 June, Trondheim, Norway

Gazetas G, Mylonakis G (1998) Seismic soil-structure interaction: new evidence and emerging issues. Geotechnical Special Publication 75, Geotechnical earthquake engineering and soil 12 dynamics III, ASCE, 2, 1119–1174

Gazetas G, Mylonakis G (2002) Kinematic pile response to vertical P wave seismic excitation. J Geotech Geoenviron Eng 128:860–867

Giuffrè A, Pinto PE (1970) Il comportamento del cemento armato per sollecitazione ciclice di forte intensità Giornale del genio civile 108(5):391–408

Goel RK, Chopra A (1997) Evaluation of bridge abutment capacity and stiffness during earthquakes. Earthq Spectra 13(1):1–23

Goel R, Chopra A (2008) Role of shear keys is seismic behaviour of bridge crossing fault-rupture zones. J Bridge Eng 13(4):398–408

GT STRUDL (2006) GT STRUDL version 29 user's reference manual, Georgia Tech. Research Corporation, Atlanta, GA

Guedes J (1997), Seismic behaviour of reinforced concrete bridges. Modelling, numerical analysis and experimental assessment. PhD thesis, University of Porto, Porto, Portugal.
Guedes J, Pegon P, Pinto AV (1994) A fibre/Timoshenko beam element in CASTEM 2000, Special Publication Nr. I.94.31 applied mechanics unit, Safety Technology Institute, Commission of the European Communities, Joint Research Centre, Ispra Establishment, Italy
Higashino M, Okamoto S (2006) Response control and seismic isolation of buildings. Taylor, Francis Ltd, London/New York
Inel M, Aschheim M (2004) Seismic design of columns of short bridges accounting for embankment flexibility. J Struct Eng ASCE 130(10):1515–1528
Isaković T, Fischinger M (1998) Engineering modelling for inelastic seismic response of RC bridge columns. Int J Eng Model 11(3/4):67–72
Isaković T, Fischinger M (2006) Higher modes in simplified inelastic seismic analysis of single column bent viaducts. Earthq Eng Struct Dyn 35(1):95–114
Jiang Y, Saiidi M (1990) 4-Spring element for cyclic response of RC columns. J Struct Eng ASCE 116(4):1018–1029
Kabeyasawa T, Shiohara H, Otani S, Aoyama H (1983) Analysis of the full-scale seven-story reinforced concrete test structure. J Fac Eng Univ Tokyo 37(2):431–478
Kanaan AE, Powell GH (1973) A general purpose computer program for dynamic analysis of planar structures. Report UBC/EERC-73/6, University of California, Berkeley
Kappos AJ (1991) Analytical prediction of the collapse earthquake for RC buildings: suggested methodology. Earthq Eng Struct Dyn 20(2):167–176
Kappos AJ, Sextos AG (2001) Effect of foundation type and compliance on seismic response of 1988 RC bridges. J Bridge Eng, ASCE 6(2):120–130
Kappos A, Potikas P, Sextos A (2007) Seismic assessment of an overpass bridge accounting for nonlinear material and soil response and varying boundary conditions. Computational methods in structural dynamics and earthquake engineering, COMPDYN 2007, Rethymnon, Greece, CD-ROM volume
Kausel E, Roesset JM (1994) Soil-structure interaction for nuclear containment structures. In: Proceedings ASCE power division specialty conference, Boulder, Colorado
Kavvadas M, Gazetas G (1993) Kinematic seismic response and bending of free-head piles in layered soil. Geotechnique 43:207–222
Kelly TE (1992) Skellerup industries, lead rubber isolation bearings: experimental properties. Holmes Consulting Group, Auckland
Kent DC, Park R (1971, July) Flexural members with confined concrete. J Struct Div, ASCE 97(ST7):1969–1990
Kotsoglou A, Pantazopoulou S (2007) Bridge–embankment interaction under transverse ground excitation. Earthq Eng Struct Dyn 36:1719–1740
Kotsoglou A, Pantazopoulou S (2009) Assessment and modelling of embankment participation in the seismic response of integral abutment bridges. Bull Earthq Eng 7:343–361
Kwon O-S, Sextos A, Elnashai A (2009) Seismic fragility of a bridge on liquefaction susceptible soil, In: 10th international conference on seismic safety and reliability, 13–17 September, Osaka, Japan
Leonhardt F (1981) From past achievements to new challenges for joints and bearings. First world conference on joints and bearings, vol. 1, American Concrete Institute, SP-70, pp 736–755
Lysmer J, Kulemeyer RL (1969) Finite dynamic model for infinite media. J Eng Mech Div ASCE 95:759–877
Mackie K, Stojadinovic B (2002) Probabilistic seismic demand model for typical highway overpass bridges. In: 12th European conference on earthquake engineering, London, UK, CD-ROM volume
Maekawa K, Pimanmas A, Okamura H (2003) Nonlinear mechanics of reinforced concrete. Spon Press, London
Makris N, Gazetas G (1992) Dynamic pile-soil-pile interaction. Part II: lateral and seismic response. Earthq Eng Struct Dyn 21(2):145–162
Mander JB, Priestley MJN, Park R (1988) Theoretical stress–strain model for confined concrete. J Struct Eng ASCE 114(ST8):1804–1826

Maragakis E, Thornton G, Saiidi M, Siddharthan R (1989) A simple non-linear model for the investigation of the effects of the gap closure at the abutment joints of short bridges. Earthq Eng Struct Dyn 18(8):1163–1178

Maroney BH, Chai YH (1994) Seismic design and retrofitting of reinforced concrete bridges. In: Proceedings of 2nd international workshop, Earthquake Commission of New Zealand, Queenstown

Martinez-Rueda JE, Elnashai AS (1997) Confined concrete model under cyclic load. Mater Struct 30(197):139–147

Mazzoni S, Mckenna F, Scott MH, Fenves GL, Jeremic B (2003) Open system for earthquake engineering simulation. (OpenSees) – command language manual. UCB, PEER, University of California, Berkeley

MCEER/ATC (2003) Recommended LFRD guidelines for the seismic design of highway bridges, Part I: Specifications, Part I: Commentary and appendices. MCEER/ATC49. MCEER Report No. MCEER-03SP03, University of Buffalo, Buffalo, NY.

Mckenna F, Fenves GL, Scott MH, Jeremić B (2000) Open system for earthquake engineering simulation. http://opensees.berkeley.edu

Megally SH, Silva PF, Seible, F (2001) Seismic response of sacrificial shear keys in bridge abutments. Report No. SSRP-2001/23, Department of Structural Engineering, University of California, San Diego

Menegotto M, Pinto PE (1973) Method of analysis for cyclically loaded R.C. plane frames including changes in geometry and non-elastic behaviour of elements under combined normal force and bending, symposium on the resistance and ultimate deformability of structures acted on by well defined repeated loads, International Association for Bridge and Structural Engineering, Zurich, Switzerland, pp 15–22

Monti G, Nuti C (1992) Nonlinear cyclic behaviour of reinforcing bars including buckling. J Struct Eng 118(12):3268–3284

Mylonakis G, Gazetas G (2000) Seismic soil-structure interaction: beneficial or detrimental? J Earthq Eng 4(3):277–301

Mylonakis G, Nikolaou A, Gazetas G (1997) Soil-pile bridge seismic interaction: kinematic and inertial effects. Part I: Soft soil, earthquake. Eng Struct Dyn 26:337–359

Mylonakis G, Papastamatiou D, Psycharis J, Mahmoud K (2001) Simplified modelling of bridge response on soft soil to nonuniform seismic excitation, ASCE J Bridge Eng 6(6):587–597

Mylonakis G, Nikolaou S, Gazetas G (2006a) Footings under seismic loading: analysis and design issues with emphasis on bridge foundations. Soil Dyn Earthq Eng 26(9):824–853

Mylonakis G, Syngros C, Gazetas G, Tazoh T (2006b) The role of soil in the collapse of 18 piers of Hanshin Expressway in the Kobe earthquake. Earthq Eng Struct Dyn 35(5):547–575

Naeim F, Kelly JM (1999) Design of seismic isolated structures: from theory to practice. Wiley, Chichester

Nagarajaiah S, Reinhorn, AM, Constantinou MC (1991) 3D-Basis: nonlinear dynamic analysis of three-dimensional base isolated structures: Part II, Technical Report NCEER- 91–0005, National Centre for Earthquake Engineering Research, State University of New York at Buffalo, Buffalo, NY

NAVFAC [Naval Facilities Engineering Command] (1982) DM-7.2, Foundations and earth structures, design manual. Department of the Navy, Alexandria

Nogami T, Konagai K, Otani J, Chen HL (1992) Nonlinear soil-pile interaction model for dynamic lateral motion. J Geotech Eng ASCE 118(1):106–116

Novak M, Mitwally H (1988) Transmitting boundary for axisymmetrical dilation problems. J Eng Mech 114(1):181–187

Paraskeva TS, Kappos AJ, Sextos AG (2006) Extension of modal pushover analysis to seismic assessment of bridges. Earthq Eng, Struct Dyn 35(11):1269–1293

Park R, Priestley M, Gill W (1982) Ductility of square-confined concrete columns. J Struct Div ASCE 108(4):929–950

Park Y, Wen Y, Ang A (1986) Random vibration of hysteretic systems under bi-directional ground motions. Earthq Eng Struct Dyn 14:543–557, Wiley, New York

Pender MJ (1993) Aseismic pile foundation design analysis. Bull N Z Natl Soc Earthq Eng 26(1):49–161

Penzien J, Watabe M (1975) Characteristics of 3D earthquake ground motions. Earthq Eng, Struct Dyn 3:365–373

Petroleum Institute (1993) Planning, designing and constructing fixed offshore platforms – working stress design, RP 2A-WSD, USA

Pinho R (2000) Shaking table testing of RC walls. ISET J Earthq Technol 37(4):119–142

Pinho R, Casarotti C, Antoniou S (2007) A comparison of single-run pushover analysis techniques for seismic assessment of bridges. Earthq Eng Struct Dyn 36(10):1347–1362

Pinto AV, Verzeletti G, Pegon P, Magonette G, Negro P, Guedes J (1996) Pseudo-dynamic testing of large-scale RC bridges, ELSA Lab, Report EUR 16378 EN

Pitilakis K, Kirtas E, Sextos A, Bolton M, Madabhushi G, Brennan A (2004) Validation by centrifuge testing of numerical simulations for soil-foundation-structure systems. In: 13th world conference on earthquake engineering, Vancouver, Paper No. 2772

Priestley MJN, Seible F, Calvi GM (1996) Seismic design and retrofit of bridges. John Wiley, Sons, Inc., New York

Renault P, Meskouris K (2004) Coupled boundary element – finite element method in soil-structure interaction analyses, advanced numerical analyses of solids and structures, and beyond. Graz University of Technology, Graz, pp 169–181

Sadrossadat-Zadeh M, Saiidi M (2007) Pre-test analytical studies of NEESR-SG 4-span bridge model using OpenSees. Centre for Civil Engineering Earthquake Research, Department of Civil Engineering, Report No. CCEER-07-3, University of Nevada, Reno, Nevada

Saiidi M (1982) Hysteresis models for reinforced concrete. J Struct Div ASCE 108(ST5): 1077–1085

Saiidi M, Ghusn G, Jiang Y (1989) Five-spring element for biaxially bent RC columns. J Struct Eng ASCE 115(2):398–416

Saiidi M, Moore R, Itani A (2001) Seismic performance of reinforced concrete bridges with unconventional configurations. ACI Struct J 98(5):717–726

Savidis SA, Bode C, Hirschauer R (2000) Three-dimensional structure-soil-structure interaction under seismic excitations with partial uplift. 12th world conference on earthquake engineering, Auckland, New Zealand.

Scott BD, Park R, Priestley MJN (1982) Stress-strain behaviour of concrete confined by overlapping hoops at low and high strain rates. ACI J 79(1):13–27

SeismoSoft (2005) SeismoStruct – A computer program for static and dynamic nonlinear analysis of framed structures [on line]. Available from URL: http://www.seismosoft.com. Accessed 30 March 2005

Sextos AG, Kappos AJ (2008) Seismic response of bridges under asynchronous excitation and comparison with EC8 design rules. Bull Earthq Eng 7(2):519–545

Sextos A, Kappos A, Pitilakis K (2003a) Inelastic dynamic analysis of RC bridges accounting for spatial variability of ground motion, site effects and soil-structure interaction phenomena. Part 2: Parametric analysis. Earthq EngStruct Dyn 32(4):629–652

Sextos A, Pitilakis K, Kappos A (2003b) Inelastic dynamic analysis of RC bridges accounting for spatial variability of ground motion, site effects and soil-structure interaction phenomena. Part 1: Methodology and analytical tools. Earthq Eng Struct Dyn 32(4):607–627

Sextos A, Kappos A, Mergos P (2004) Effect of soil-structure interaction and spatial variability of ground motion on irregular bridges: the case of the Krystallopigi bridge, 13th world conference on earthquake engineering, Paper No. 2298, Vancouver, August 2004

Sextos A, Mackie K, Stojadinovic B, Taskari O (2008) Simplified P-y relationships for modelling embankment abutment systems of typical California bridges, 14th world conference on earthquake engineering, Beijing, China

Sextos A, Taskari O (2009) Single and multi-platform simulation of linear and non-linear bridge-soil systems, chapter in Coupled Site and Soil-Structure Interaction Effects with Application to Seismic Risk Mitigation, Springer

Shamsabadi A (2006) Three-dimensional nonlinear seismic soil-abutment-foundation structure interaction analysis of skewed bridges. PhD dissertation, Department of Civil and Environmental Engineering, University of Southern California, Los Angeles, CA

Shamsabadi A, Rollins KM, Kapuskar M (2007) Nonlinear soil–abutment–bridge structure interaction for seismic performance-based design. J Geotech Geoenviron Eng, ASCE 13(6):707–714

Shinozuka M, Saxena V, Deodatis G (2000) Effect of spatial variation of earthquake ground motion on highway structures. Technical Report, MCEER-00-0013

Siddharthan R, El-Gamal M, Maragakis E (1997) Stiffnesses of abutments on spread footings with cohesionless backfill. Can Geotech J 34:686–697

Silva PF, Megally S, Seible F (2003) Seismic performance of sacrificial interior shear keys. ACI Struct J 100(2):177–187

Skinner RI, Robinson H, McVerry GH (1993) An introduction to seismic isolation. Wiley, Chichester

Spacone E, Ciampi V, Filippou FC (1992) A beam element for seismic damage analysis, Report EERC 92–07, Earthquake Engineering Research Center. University of California, Berkeley

Spyrakos C, Ioannidis G (2003) Seismic behaviour of a post-tensioned integral bridge including soil–structure interaction (SSI). Soil Dyn Earthq Eng 23:53–63

Sullivan T, Pinho R, Pavese A (2004) An introduction to structural testing techniques in earthquake engineering, Educational Report ROSE 2004/01, IUSS Press, Pavia

Takeda T, Sozen MA, Nielsen NN (1970) Reinforced concrete response to simulated earthquakes. J Struct Mech Div ASCE ST12:96

Taucer F, Spacone E, Filippou FC (1991) A fiber beam-column element for seismic response analysis of RC structures, Report EERC 91–17, Earthquake Engineering Research Center, University of California. Berkeley

Taylor AW, Lin AN, Martin JW (1992) Performance of elastomers in isolation bearings: a literature review. Earthq Spectra 8(2):279–304

Tokimatsu K (1999) Performance of pile foundations in laterally spreading soils. Proc 2nd Int Conf Earthq Geotechnical Eng, Lisbon, Portugal 3:957–964

Vesic A (1961) Beams on elastic foundations. Proc 5th Int Conf Soil Mech Foundation Eng, Paris 1:845–850

Vulcano A, Bertero VV, Caloti V (1989) Analytical modelling of RC structural walls. Proc 9th WCEE, Tokyo-Kyoto, Maruzen 6:41–46

Wang M-L, Shah SP (1987) Reinforced concrete hysteresis model based on the damage concept. Earthq Eng Struct Dyn 15:993–1003

Wolf JP (1994) Foundation vibration analysis using simple physical models. Prentice Hall, Englewood Cliffs

Zayas V, Low S (1990) A simple pendulum technique for achieving seismic isolation. Earthq Spectra 6(2):317–333

Zhang J, Makris N (2001) Seismic response of highway overcrossings including soil–structure interaction. PEER Report 2001/02, University of California, Berkeley

Zhang J, Makris N (2002) Kinematic response functions and dynamic stiffnesses of bridge embankments. Earthq Eng Struct Dyn 31:1933–1966

Chapter 3
Methods for Inelastic Analysis of Bridges

M. Nuray Aydınoğlu, Matej Fischinger, Tatjana Isaković, Andreas J. Kappos, and Rui Pinho

3.1 Introduction

Nonlinear response analysis procedures currently used for performance-based assessment and design of bridges (and buildings) can be broadly classified into two groups, namely,

(a) *Nonlinear response history analysis (NRHA) procedure*, and
(b) *Practice-oriented nonlinear analysis procedures* based on the so-called *push-over analysis,*

which are presented in the following sections.

The NRHA procedure is briefly presented first (Sect. 3.2), since it is, by definition, the most accurate method, as it accounts for both the inelastic response characteristics and the dynamic effects in the structure, while the input for the analysis is the ground motion itself (typically a set of accelerograms). This statement should not be

M. Nuray Aydınoğlu (✉)
Department of Earthquake Engineering, Kandilli Observatory and Earthquake Research Institute, Boğaziçi University, 34684 Çengelköy – Istanbul, Turkey
e-mail: aydinogn@boun.edu.tr

M. Fischinger • T. Isaković
Faculty of Civil and Geodetic Engineering, University of Ljubljana, Jamova 2, 1000 Ljubljana, Slovenia
e-mail: matej.fischinger@ikpir.fgg.uni-lj.si; tisak@ikpir.fgg.uni-lj.si

A.J. Kappos
Department of Civil Engineering, Aristotle University of Thessaloniki, 54124 Thessaloniki, Greece
e-mail: ajkap@civil.auth.gr

R. Pinho
Department of Structural Mechanics, University of Pavia, Via Ferrata 1, 27100 Pavia, Italy
e-mail: rui.pinho@unipv.it

A.J. Kappos et al. (eds.), *Seismic Design and Assessment of Bridges: Inelastic Methods of Analysis and Case Studies*, Geotechnical, Geological and Earthquake Engineering 21, DOI 10.1007/978-94-007-3943-7_3, © Springer Science+Business Media Dordrecht 2012 construed as implying that all inelastic dynamic analysis models describe accurately the seismic response of a given bridge; the reliability of any model depends on the way the geometry, as well as the stiffness characteristics of the structure, and the way it is supported on the ground, are modelled. Hence in the remainder of this document, comparisons of approximate (static or 'quasi-dynamic') methods with the 'reference method' i.e. NRHA should be understood as evaluation of the relative accuracy of each procedure applied to a model having essentially the same geometry and initial stiffness and boundary condition characteristics.

The next section (Sect. 3.3) includes some general considerations on the nonlinear static (pushover) analysis methods and a brief history thereof, while Sect. 3.4 presents in more detail the single-mode pushover analysis procedures. The focus in this section is on the 'N2' method, which is essentially the one adopted in the current European code, Eurocode 8. This establishes an appropriate reference for other methods (currently having the status of research proposals) to be assessed, i.e. permits the evaluation of the possible advantages (or otherwise) the new, generally more involved, proposals offer, compared to the current code method. Of course, the accuracy of all procedures (including the code-prescribed inelastic static methods) are eventually judged in the light of NRHA (see also Chap. 4).

The section that follows (Sect. 3.5) is arguably the most important part of this chapter, wherein different recent proposals for approximate inelastic analysis methods are presented in sufficient detail and in a reasonably uniform way, i.e. clearly stating the assumptions and simplifications involved and describing the steps involved in the application of each method. Comprehensive case-studies, involving the application of several different methods (including NRHA) to the same bridges and evaluation of the results are found in Chap. 4.

3.2 Nonlinear Response History Analysis (NRHA) Procedure

The NRHA procedure involves direct integration, in the time domain, of the equations of motion for a given set of ground motion inputs, and capitalises on the fact that modelling the nonlinear cyclic behaviour of commonly used materials, i.e., reinforced concrete and structural steel is now well established.

Selection and scaling of ground motion records are considered to be one of the most complex issues in practice. Generally, seven sets of bi-directional records complying with certain scaling criteria are required by the codes, under which design quantities are approximated as median values. Occasionally three ground motion sets are allowed provided that peak values are considered for design.

Typically, a "lumped plasticity approach" based on conventional plastic-hinge hypothesis is adopted in practical applications, taking advantage of the relatively modest computational demand of such analysis typology (as compared with the fibre analysis approach described below). However, selection and calibration of appropriate hysteretic models (Fig. 3.1) for the different structural elements, together with appropriate plastic hinge lengths, is not always straightforward,

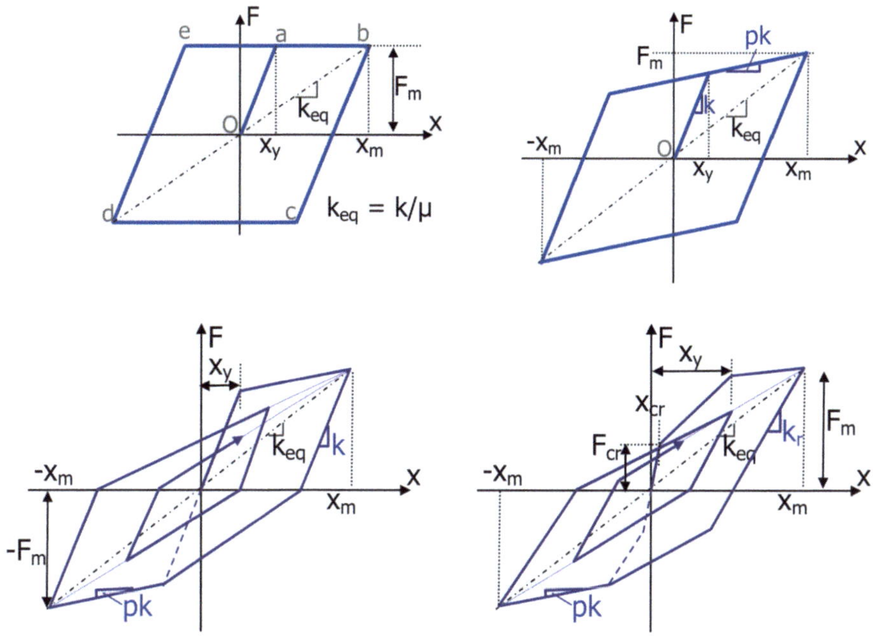

Fig. 3.1 Examples of hysteresis rules used for structural members

since the analyst needs to juggle with parameters such as axial load ratio, solid vs. hollow section type, shear span ratio, reinforcement percentage, confinement level, etc., all of which play a decisive role in the hysteretic response of structural members. Such difficulties have in the past limited the spread of NHRA application in design offices.

During the last decade or so, however, a more refined "distributed plasticity approach" based on fibre elements has emerged as a viable alternative. Whilst a detailed discussion on the features of distributed inelasticity modelling, together with a description of its historical evolution, can be found e.g. in Filippou and Fenves (2004) or Fragiadakis and Papadrakakis (2008), herein it is simply mentioned that in fibre modelling each fibre is associated with a uniaxial stress–strain relationship; the sectional stress–strain state of beam-column elements is then obtained through the integration of the nonlinear uniaxial stress–strain response of the individual fibres in which the section has been subdivided (the discretisation of a typical reinforced concrete cross-section is depicted, as an example in Fig. 3.2). Such approach thus features the following assets: no requirement of a prior moment-curvature analysis of members, no need to introduce any element hysteretic response (as it is implicitly defined by the material constitutive models), direct modelling of axial load-bending moment interaction (both on strength and stiffness), straightforward representation of biaxial loading, and interaction between flexural strength in orthogonal directions. On the other hand, such models generally overestimate the initial stiffness of RC members since they ignore existing cracking

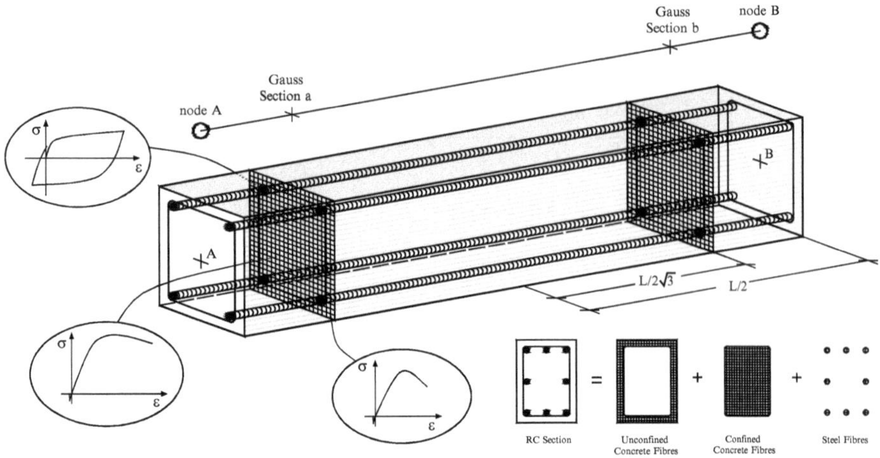

Fig. 3.2 Example of section fibre discretisation

(that typically exists due to environmental actions like shrinkage and/or due to past loading history), while they cannot model inelastic shear and bond-slip effects (without introducing additional features, such as end springs, which are not available in most of the available software packages in this category).

Distributed inelasticity frame elements can be implemented with two different finite element (FE) formulations: the classical displacement-based (DB) ones (e.g. Hellesland and Scordelis 1981; Mari and Scordelis 1984), and the more recent force-based (FB) formulations (e.g. Spacone et al. 1996; Neuenhofer and Filippou 1997). In a DB approach the displacement field is imposed, whilst in a FB element equilibrium is strictly satisfied and no restraints are placed to the development of inelastic deformations throughout the member; see e.g. Alemdar and White (2005) and Freitas et al. (1999) for further discussion. This latter formulation is thus recently gaining ground as preferred fibre modelling approach, since it does tend to provide more accurate results without the need for mesh refinement (provided that 4–5 integration sections are used throughout the element's length, which is typical).

A word of caution regarding the undertaking of NHRA refers to the definition of viscous damping to represent energy dissipation sources that are not explicitly included in the hysteretic models of structural elements. Traditionally, such modest energy dissipation sources have been considered through the use of Rayleigh damping e.g. Clough and Penzien (1993), Chopra (2001) with equivalent viscous damping values varying from 1% to 8%, depending on structural type, materials used, non-structural elements, period and magnitude of vibration, mode of vibration being considered, etc. (e.g. Wakabayashi 1986). However, some authors (e.g. Wilson 2001) strongly suggest for such equivalent modelling to be avoided altogether, whilst others (Priestley and Grant 2005; Hall 2006) advise its employment but not by means of Rayleigh damping, which is proportional to both mass and stiffness, but rather through the use of stiffness-proportional damping only;

as discussed by Pegon (1996), Wilson (2001), Abbasi et al. (2004) and Hall (2006), amongst others, if a given structure is "insensitive" to rigid body motion, mass-proportional damping will generate spurious (hence unrealistic) energy dissipation. The stiffness-proportional damping modelling approach may then be further sub-divided in initial stiffness-proportional damping and tangent stiffness-proportional damping, the latter having been described by Priestley and Grant (2005).

Finally, it is strongly recommended that NRHA be always accompanied by the undertaking of simpler but more easily checkable nonlinear static analyses, which will give analysts the possibility of spot-checking the complex and somewhat overwhelming outputs of dynamic runs, by comparing e.g. the peak displacement/force response values obtained in the latter with the estimates provided by the former. Hand calculations may also be used to estimate member capacities and ensure that these are being respected in NHRA (spurious vibration modes, originating in inadequate meshing and/or choice of integration parameters, may indeed lead to the attainment of unrealistic high dynamic response values).

A number of appropriate software packages is available for practical applications where nonlinear hysteretic characteristics of elements and/or components can be fully described including stiffness degradation, strength deterioration, pinching, etc.

3.3 Practice-Oriented Nonlinear Analysis Procedures Based on Pushover Analysis

3.3.1 General

In view of the lack of sufficient experience and professional training, practicing engineers find themselves unprepared to evaluate and interpret the results of nonlinear response history analysis. It appears, at least for the time being, they tend more favourably to use the so-called *practice-oriented nonlinear analysis procedures*, which are mainly based on pushover analysis. Pushover-based methods are considered more user-friendly and relatively easy-to-understand. Moreover, such methods have a great advantage in specifying the seismic input, i.e., they employ the familiar elastic response spectra instead of selecting and scaling ground motion histories. Modelling the nonlinear behaviour essentially follows NRHA without any requirement for hysteretic properties.

Essentially, all pushover analysis procedures can be interpreted as approximate extensions of linear *Response Spectrum Analysis (RSA)* method to the nonlinear response analysis with varying degrees of sophistication. In this respect *Nonlinear Static Procedure – NSP* (ATC 1996; ASCE 2000) or the so-called *N2 Method* (Fajfar and Fischinger 1987; CEN 2004; Fajfar 2007) may be looked upon as a *single-mode inelastic response spectrum analysis* procedure where the peak response may be obtained through a nonlinear analysis of an equivalent modal single-degree-of-freedom (SDOF) system. In practical applications however, modal peak response is routinely estimated through *inelastic displacement spectrum* (Fajfar 1999; CEN 2004). On the other hand, a number of multi-mode pushover procedures have been developed in the last decade for estimating the seismic demand, which may be considered as approximate versions of *multi-mode inelastic response spectrum analysis*. Typical examples are the Modal Pushover Analysis (MPA) by Chopra and Goel (2002) and the Incremental Response Spectrum Analysis (IRSA) by Aydınoğlu (2003, 2004), which are described in detail in Sect. 3.5.

3.3.2 Historical vs. Contemporary Implementation of Pushover Analysis

From a historical perspective, pushover analysis has always been understood and implemented as a nonlinear *capacity estimation tool* and is often called *capacity analysis*. The nonlinear structure is monotonically pushed by a set of forces with an invariant distribution until a predefined displacement limit at a given location (say, lateral displacement limit at the roof level of a building) is attained. Such predefined displacement limit is generally termed *target displacement*. The structure may be further pushed up to the collapse limit in order to estimate its ultimate deformation and load carrying capacities. It is for this reason that pushover analysis has been also called *collapse analysis*.

Note that in view of performance-based seismic assessment and design requirements, such a definition of pushover analysis is not sufficient. According to a new concept first introduced by Freeman et al. (1975) and Fajfar and Fischinger (1987), which was subsequently adopted in ATC 40 (ATC 1996); FEMA 273 (FEMA 1997); FEMA 356 (ASCE 2000) and Eurocode 8 (CEN 2004), pushover analysis with its above-given historical definition represents only the first stage of a two-stage nonlinear static procedure, where it simply provides the nonlinear capacity curve (pushover curve) of the structural system. The ultimate goal is the estimation of the peak response, i.e. the seismic demand, which is achieved in the second stage through the nonlinear response history analysis of an equivalent single-degree-of-freedom (SDOF) system under a given earthquake or, preferably, through an inelastic displacement spectrum. In this sense the term *pushover analysis* now includes as well the estimation of the so-called *target displacement* as a seismic demand parameter. Thus according to this broader definition, pushover analysis is not only a capacity estimation tool, but at the same time it is a *demand estimation tool*. Fajfar and Fischinger (1987, 1988) judiciously coined the name *N2 Method* based on this two-stage concept (see Sect. 3.4.1 below).

It should be noted that several multi-mode pushover procedures that have been developed in the last decade (see Sect. 3.5.3) fall into the class of *capacity estimation tools*, i.e. they are unable to estimate the seismic demand under a given earthquake action. Multi-mode pushover analysis procedures that are able to estimate the seismic demand are presented in Sect. 3.5 below.

3.4 Single-mode Pushover Analysis Procedures

The procedures falling within this category are characterised by an invariant or adaptive distribution of static-equivalent seismic loads, where the former can be either those of the dominant mode in the earthquake direction considered (the fundamental mode in a 2D analysis) or a code-specified distribution ('triangular', 'uniform'). The target displacement that represents the seismic demand can be defined in a number of different ways, some more involved than others. It is beyond the scope of this book to review and discuss all available procedures; this is done in a number of studies, a recent one being ATC-55 (FEMA 2005). It will be noted, though, that code-type documents usually adopt one of the following methods:

- The "*Displacement Coefficient Method*" (FEMA 1997; MPWS 2007; OASP 2009), wherein the elastic spectral displacement of the fundamental mode is modified through a number of coefficients accounting for nonlinear effects in the equivalent SDOF system and its conversion to MDOF system.
- The "*Capacity Spectrum Method*" (Freeman 1998), wherein the seismic demand is expressed either in the form of elastic spectra for high equivalent damping values (estimated from the ductility demand in the structure) as originally proposed by Freeman et al. (1975) and further developed in ATC-40 (1996), or in the form of inelastic demand spectra (Fajfar 1999; Chopra and Goel 1999; CEN 2004).

For the reasons explained in the previous section, the N2 method is presented in the following as a typical representative of this category. It has to be noted here that the idea of representing a structure (such as a RC building) by an equivalent single-degree-of-freedom (SDOF) oscillator whose characteristics are defined from a nonlinear static analysis of the corresponding multi-degree-of-freedom (MDOF) structure was first explored by Saiidi and Sozen (1981), who extended previous work by Biggs (1964) on equivalent SDOF systems.

3.4.1 Single-mode Pushover Analysis Procedure with Invariant Load Patterns: The N2 Method

The N2 method (Fajfar and Fischinger 1987; Fajfar 2007) is a nonlinear static (pushover) method, which is also included into the standard Eurocode 8 (CEN 2004, 2005). The name of the method describes its basic features. The N stands for the nonlinear analysis and 2 means two types of the analysis on two different analytical models. Namely, the method is a combination of the static nonlinear analysis of the MDOF model and nonlinear dynamic analysis of equivalent SDOF model. The nonlinear static analysis of the MDOF model is used to define the properties of the structure, which are further used to define the equivalent SDOF model on which the dynamic analysis is performed.

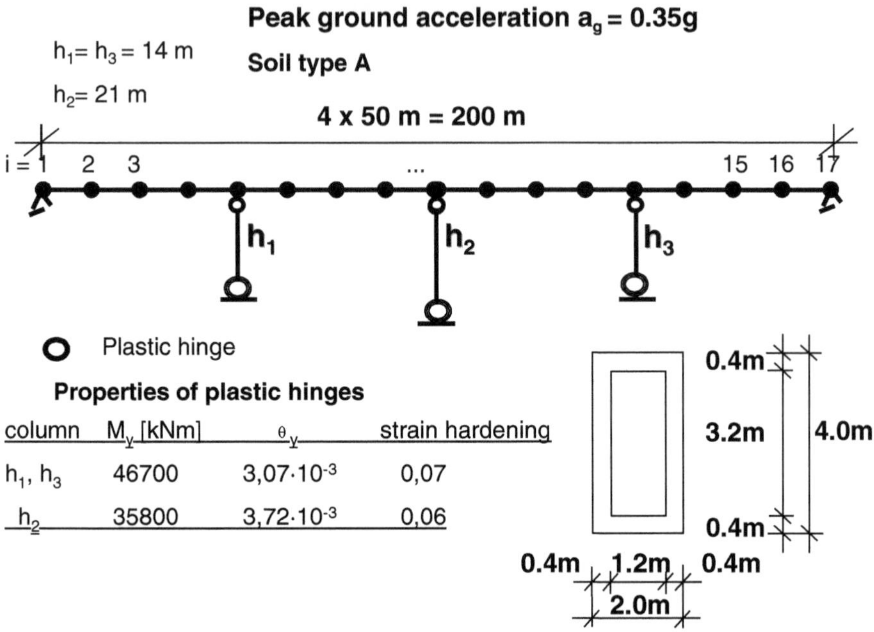

Fig. 3.3 An example of a bridge, used to illustrate the steps of the N2 method

In general, the N2 method makes the nonlinear seismic analysis of structures straightforward, since the dynamic analysis can be performed on the simplified model using the nonlinear response spectra. The dynamic analysis becomes so simple that special software for the nonlinear dynamic analysis is not needed. The nonlinear response spectra can be determined from the standard elastic response spectra. This is one of the main advantages compared to the nonlinear response history analysis (NRHA), since the seismic load can be simply standardized.

There are several variants of the N2 method defined for different types of structures. The variant of the N2 method, which is deemed to be the most appropriate for bridges, is described in the next chapter. The major steps are illustrated by a simple numerical example. Then some specifics related to the application of the method for bridges are discussed and summarized.

3.4.1.1 Step-By-Step Implementation of the N2 Method for Bridges

The N2 method is described step-by-step on the example of the bridge presented in Fig. 3.3. The analysis in the transverse direction is described.

Step 1: *Definition of the MDOF model of the bridge, which is used for the nonlinear static analysis.*

The model of the bridge, which is appropriate for the nonlinear static analysis, should be defined first. A similar model as that suitable for the elastic analysis, can be used. This model should be enhanced by additional parameters that define the nonlinear force-deformation relationship for each structural element. The simplest model includes standard beam-column elements with plastic hinges at both nodes. Description of several suitable models can be found in Chap. 2 – Modelling of Bridges for Inelastic Analysis.

In bridges the nonlinear response is usually limited to columns, while it is usually supposed that the superstructure will remain in the elastic range. Thus, in the bridge presented in Fig. 3.3, the superstructure was modelled by standard elastic beam-column elements. When modelling columns, it was assumed that the nonlinear deformation could occur only at the connections with footings (plastic hinges were added at the basement of the columns). At the top of the columns a pinned connection with the superstructure was taken into account. Properties of the plastic hinges in columns of the bridge, which is used to illustrate the N2 method, are presented in Fig. 3.3.

Step 2: *Nonlinear static analysis – pushover analysis, used to define the relationship between seismic forces and displacements of the structure.*

In the nonlinear static (pushover) analysis the model defined in the 1st step is subjected to horizontal lateral load. The intensity of this load, applied along the superstructure, is gradually increased and the deformations (displacements) of the structure are monitored. In this way a characteristic non-linear force-displacement relationship is determined. The relationship between total base shear (in columns and the abutments) and the maximum displacement of the superstructure is typically monitored. In many bridges the maximum displacement will occur in the centre or close to the centre of the superstructure (see the notes in the next section).

The relationship between the base shear and maximum displacement depends mainly on the distribution of the lateral (horizontal) load, applied along the superstructure. It is typically assumed that the applied forces are proportional to the normalized displacements of the superstructure (displacement shape) Φ. When the analysis is performed in one direction of the bridge (e.g. in the transverse direction in the case of the illustrative example) this load can be calculated as:

$$F_{\mathrm{i}} = p\, m_{\mathrm{i}}\, \Phi_{\mathrm{i}} \tag{3.1}$$

F_{i} is the force at the position i in the bridge, m_{i} is the mass at the position i, Φ_{i} is assumed normalized displacement at position i, and p factor, which controls the magnitude of the lateral load (see Table 3.1).

The actual displacement shape Φ is not known in advance. It should be assumed. In the bridge, presented in Fig. 3.3, parabolic displacement shape was taken into account (see the comments in the next section).

Forces F_{i} are gradually increased and the maximum displacement of the superstructure is monitored. In this way the relationship, presented in Fig. 3.4a is obtained.

Step 3: *Definition of the equivalent single-degree-of-freedom (SDOF) model*

In the N2 method, seismic demand is usually determined by using response spectra. Inelastic behaviour is taken into account explicitly. Consequently, the structure

Table 3.1 Calculation of mass m* of SDOF system and factor Γ

i	Mass m_i [t]	Φ_i	$F_i = m_i\Phi_i$	$m_i\Phi_i^2$
1	127.4	0	0	0
2	254.8	0.234	59.7	14.0
3	254.8	0.438	111.5	48.8
4	254.8	0.609	155.3	94.6
5	329.0	0.750	246.8	185.1
6	254.8	0.859	219	188.2
7	254.8	0.938	238.9	224.0
8	254.8	0.984	250.8	246.9
9	366.1	1.000	366.1	366.1
10	254.8	0.984	250.8	246.9
11	254.8	0.938	238.9	224.0
12	254.8	0.859	219.0	188.2
13	329	0.750	246.8	185.1
14	254.8	0.609	155.3	94.6
15	254.8	0.438	111.5	48.8
16	254.8	0.234	59.7	14.0
17	127.4	0	0	0

m_i Masses, Φ_i normalized displacements, F_i assumed lateral load

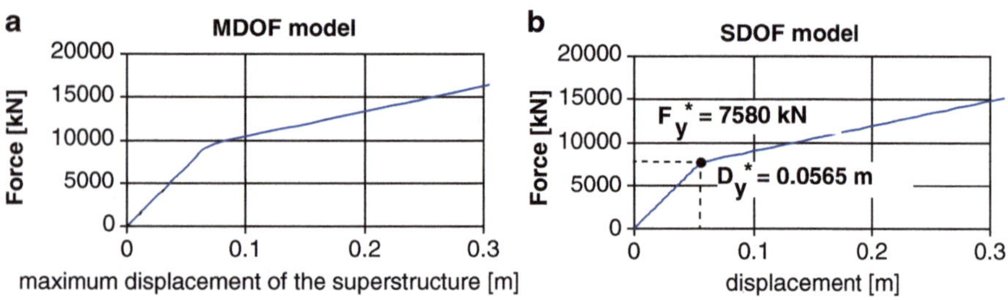

Fig. 3.4 (**a**) Relationship base shear-maximum displacement of the superstructure of the bridge, presented in Fig. 3.1; (**b**) capacity curve

should, in principle, be modelled as an SDOF system. Different procedures have been used to determine the characteristics of an equivalent SDOF system. One of them is summarized below.

The equivalent SDOF model is defined based on its stiffness k^*, the mass m^* and the period T^*. The stiffness k^* of the SDOF model (Fajfar 2007) is calculated based on the relationship of the base-shear and displacements, defined in the 2nd step (Fig. 3.4b). These forces and displacements are divided by factor Γ:

$$\Gamma = \frac{m^*}{\sum m_i \Phi_i^2} \tag{3.2}$$

$$m^* = \sum m_i \Phi_i \tag{3.3}$$

where m_i are masses of the MDOF model (see Fig. 3.3), Φ_i are assumed normalized displacements, and m^* mass of the equivalent SDOF model. The calculation procedure for m^* and Γ is illustrated in Table 3.1. For the bridge presented in Fig. 3.1, $m^* = 2930$ and $\Gamma = 1.24$.
The stiffness of the SDOF system k^* has a value:

$$k^* = \frac{F_y^*}{D_y^*} \tag{3.4}$$

F_y^* and D_y^* are the yielding force and displacement (see Fig. 3.4b). In the numerical example they have values $F_y^* = 7580$ kN, $D_y^* = 5.65$ cm. The stiffness of the structure has a value of $k^* = 134200$ kN/m.
The period of the SDOF system is calculated as:

$$T^* = 2\pi\sqrt{\frac{m^*}{k^*}} \tag{3.5}$$

and has a value of $T^* = 0.928$ s.

Step 4: *Nonlinear dynamic analysis of SDOF model*

Based on the period T^* of the SDOF system (calculated in the previous step) and based on the nonlinear response spectrum (see Fajfar 2000, 2007) the seismic demand for the SDOF model is calculated. When the displacement demand S_d in the medium- and long-period bridges ($T^* \geq T_C$; T_C is characteristic period in the elastic spectrum defined in the standard Eurocode 8/1 – EC8/1; CEN 2004) is calculated, the equal displacement rule can be used. When the displacement demand S_d is calculated based on the nonlinear spectra, determined based on the EC8/1 elastic acceleration spectrum, this displacement has a value:

$$\begin{aligned} S_d &= \frac{a_g S \cdot 2.5 \cdot T_C T^*}{4 \cdot \pi^2} \quad & T_C \leq T^* \leq T_D \\ S_d &= \frac{a_g S \cdot 2.5 \cdot T_C T_D}{4 \cdot \pi^2} \quad & T^* \geq T_D \end{aligned} \tag{3.6}$$

a_g is the ground acceleration, S the soil factor, T_C is the period representing the upper limit of the region of constant spectral accelerations, T_D is the period representing the beginning of the region of constant displacement response.
For the short period bridges ($T^* < T_C$) the procedure used to define the S_d is somewhat more complex (Fajfar 2007). Since the bridges are typically structures with relatively long periods, the equations presented above will be applicable in the majority of cases. In the case of the bridge, presented in Fig. 3.3, for a spectrum with $\alpha_g = 0.35$ g, $S = 1.0$, $T_C = 0.4$ s, and $T_D = 2.0$ s, the displacement demand S_d of equivalent SDOF structure has a value of:

$$S_d = \frac{0.35 \cdot 9.81 \cdot 1.0 \cdot 2.5 \cdot 0.4 \cdot 0.928}{4 \cdot \pi^2} = 0.0807m$$

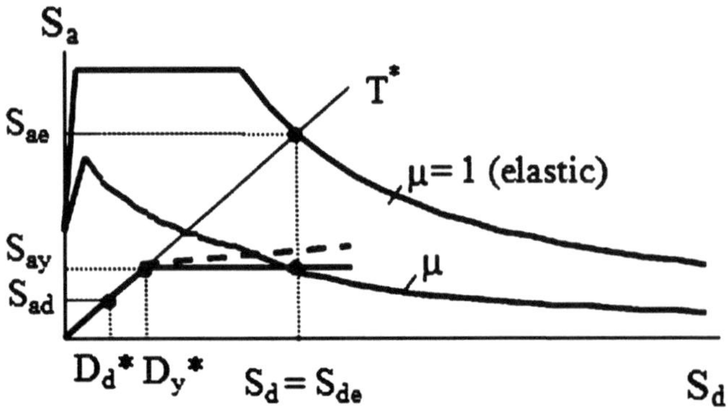

Fig. 3.5 Elastic and inelastic demand spectra versus capacity diagram

When the nonlinear spectra are determined based on some other type of spectrum, the procedure described by Fajfar (2007), can be used to obtain the S_d.

The determination of the seismic demand for the equivalent SDOF system can be presented graphically as well. For medium and long period structures the graphical interpretation of the procedure, used to define the seismic demand for the SDOF system, is presented in Fig. 3.5. For short-period structures see Fajfar (2000). Both, the demand spectra and the capacity diagram have been plotted in the same graph in S_a–S_d format (Fajfar 2000, 2007). The intersection of the radial line corresponding to the elastic period T^* of the idealized bilinear system with the elastic demand spectrum defines the acceleration demand (strength), required for elastic behaviour S_{ae}, and the corresponding elastic displacement demand S_{de}. S_{ay} represents both the acceleration demand and the capacity of the inelastic system expressed in terms of the yield acceleration.

The reduction factor R_μ can be determined as the ratio between the accelerations corresponding to the elastic and inelastic systems:

$$R_\mu = \frac{S_{ae}(T^*)}{S_{ay}} \tag{3.7}$$

Note that R_μ is not the same as the reduction (behaviour, response modification) factor R used in seismic codes. The code reduction factor R takes into account both energy dissipation and the so-called overstrength. The design acceleration S_{ad} is typically smaller than the yield acceleration S_{ay}.

If the elastic period T^* is larger than or equal to T_C, the inelastic displacement demand S_d is equal to the elastic displacement demand S_{de}.

$$S_{de} = \frac{T^{*2}}{4\pi^2} S_{ae}(T^*) \tag{3.8}$$

From triangles in Fig. 3.5 it follows that the ductility demand, defined as $\mu = S_d/D_y^*$ is equal to R_μ.

$$S_d = S_{de}(T*) \qquad T* \geq T_C; \qquad \mu = R_\mu \tag{3.9}$$

The inelastic demand in terms of accelerations and displacements corresponds to the intersection point of the capacity diagram with the demand spectrum corresponding to the ductility demand μ, provided that the post-yield stiffness in the capacity diagram is zero. At this point, the ductility factor determined from the capacity diagram and the ductility factor associated with the intersecting demand spectrum are equal. In the case of a post-yield stiffness different from zero, the intersection point is determined with the horizontal line through the yield acceleration rather than with the capacity diagram.
In the case of the numerical example $S_{ay} = 2.59$ m/s^2, $S_{ae} = 3.70$ m/s^2, $\mu = R_\mu = 1.43$.

Step 5: *Analysis of the MDOF system in order to determine the seismic demand of the bridge components*

Based on the displacement S_d of the SDOF system, the maximum expected displacement D of the MDOF system is calculated using the factor Γ (see the 3rd step):

$$D = S_d \cdot \Gamma \tag{3.10}$$

In the case of the numerical example the maximum displacement of the MDOF system has a value of $D = 8.1 \cdot 1.24 = 10\,cm$.
Then the static nonlinear analysis of the MDOF system, performed in the 2nd step is repeated. Structure is pushed with the lateral load defined in the 2nd step with the intensity that corresponds to the maximum displacement D. Then the deformations of the structure and the local seismic demand in each element could be analyzed in more detail.

3.4.1.2 Specifics in the Application of the N2 Method for Bridge Analysis

The N2 method was initially developed for the analysis of buildings. There is an impression that it can be easily used for the analysis of bridges as well, since the structural system of bridges seems to be simpler than that of buildings. However, in spite of the simplicity of the structure, the response of bridges, particularly in their transverse direction, is quite often very complex and very different from that of buildings. Consequently, the variations of the N2 method are needed, taking into account the specifics of the bridge seismic response. Some of these variations are included into the Eurocode 8/2 (CEN 2005). However, several researches (Isaković and Fischinger 2006; Isaković et al. 2008) proved that some of them are not always appropriate.

The modifications of the N2 method, when it is used for the analysis of bridges, are related to:

1. The choice of the point where the displacements are monitored in order to obtain the force-displacement relationship (see 2nd step of the method),
2. The distribution of the horizontal (inertial) forces along the superstructure (see 2nd step of the method)
3. Idealization of the capacity curve and calculation of the yielding force F_y^* and yielding displacement D_y^* (see 3rd step of the method).

The N2 method assumes that the response of the structure is governed by one predominant mode which does not significantly change when the structure is subjected to seismic load of different intensities. These assumptions make the analyses very simple; however they also define the limitations of the method. These limitations are described and illustrated in Chap. 4, where several case studies of different bridges are presented.

Choice of the Monitoring Point

One of the crucial steps in the application of the N2 method is the static nonlinear analysis of the MDOF system. Based on this analysis the base shear-displacement relationship is determined, which is further used to define the properties of the equivalent SDOF system. To define this relationship an appropriate position at the superstructure should be chosen to monitor the displacements. According to the authors' opinion the monitoring point in bridges should be the position of the maximum displacement of the superstructure. This choice is clearer when the bridge is considered as an elastically supported beam (supported by columns). When the maximum displacement is monitored, the stiffness of the superstructure, supported by columns could be adequately estimated.

In some bridges the position of the maximum displacement can significantly vary when the intensity of the seismic load is changed. According to Isaković and Fischinger (2006) the force-displacement relationship should be constructed monitoring the current maximum displacement, even when its position is changing.

In some bridges (e.g. those with central stiff columns in the central part of the bridge), the monitoring point defined in this way can be quite different from that proposed in the EC8/2 (Annex H). In the standard, the centre of the mass of the bridge is proposed as the monitoring point. The centre of mass is usually in the centre or close to the centre of the superstructure. This position usually coincides with the position of the maximum displacement in bridges, which are not supported by short and stiff central columns.

The Distribution of the Lateral Forces Along the Superstructure

The distribution of the inertial forces (lateral load) should be assumed before the nonlinear static analysis is performed. In the Annex H of Standard EC8/2 two

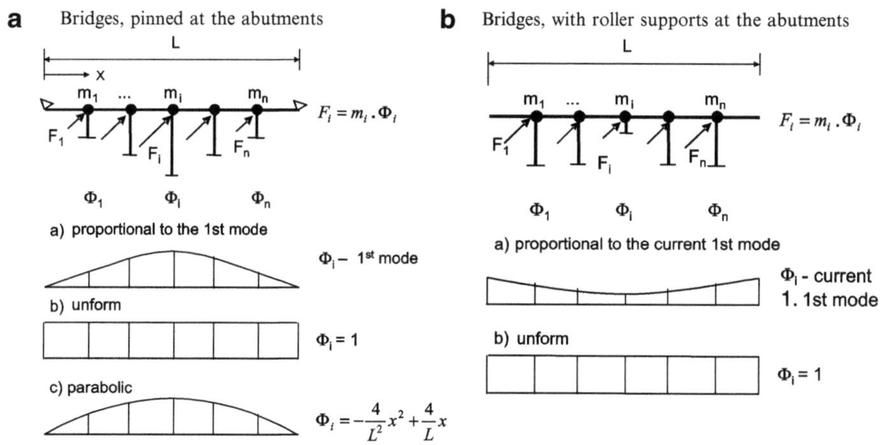

Fig. 3.6 Appropriate distributions of the lateral load

possible distributions are proposed: (a) distribution proportional to the 1st mode of the bridge in the elastic range, (b) uniform distribution (see Fig. 3.6(A)a, b). The first distribution can be defined based on the simple modal analysis with some of the standard programs for elastic modal analysis.

Isaković and Fischinger (2006) found that the parabolic distribution (Fig. 3.6(A) c) was also appropriate for bridges that are pinned at the abutments. This distribution is simpler to define than that proportional to the 1st mode. Using the parabolic distribution, in many bridges the response can be better estimated than in the case of the uniform distribution (Isaković and Fischinger 2006; Isaković et al. 2008).

In bridges with roller supports at the abutments, the uniform distribution as well as that proportional to the most important mode can be used (see Fig. 3.6(B)). Better results can be obtained if the shape of the most important mode is determined for the particular seismic intensity (Isaković and Fischinger 2006). In some bridges (e.g. bridges with short stiff central columns) this solution demands iterations, since the most important mode can considerably change with the intensity of the load.

In general it is recommended to use two different distributions of inertial forces and to take into account the envelope of the related response (as proposed in the standard).

The distribution of the lateral load does not influence only the shape of the displacement line of the superstructure but also the value of the maximum displacement. In typical bridges (without stiff central columns) pinned at the abutments, the distribution proportional to the 1st mode yields the largest displacements in the central part of the bridge, while the uniform distribution gives the largest displacements in the regions close to the abutments (Isaković et al. 2008). The parabolic distribution usually results in a deflection line in-between previously described distributions.

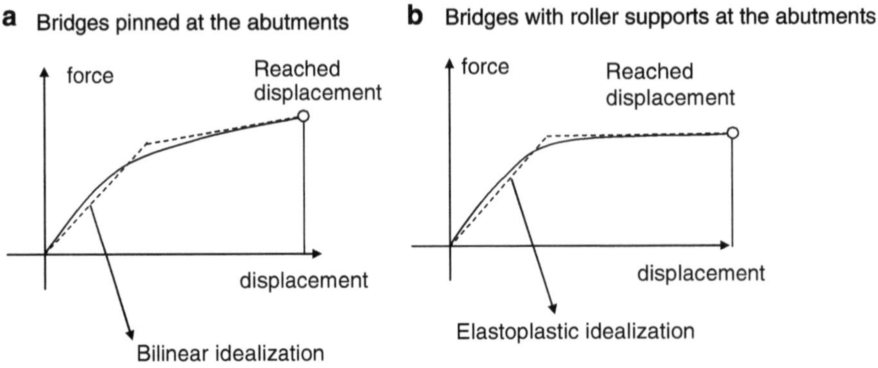

Fig. 3.7 Idealization of the capacity curve

Idealization of the Force-Displacement Curve

In the third step of the N2 method the stiffness of the SDOF system is determined based on the estimated value of the yielding force F_y^* and yielding displacement D_y^*. In the regular bridges, (e.g. that presented in Fig. 3.3), the yielding point is clear, since all columns yield almost at the same time. For other bridge configurations and different nonlinear models of columns, the capacity curve (force-displacement relationship of SDOF model) will be similar to that presented in Fig. 3.7. In such cases the capacity curve should be idealized for the yielding point to be obtained.

Originally an elastoplastic idealization of the capacity curve was proposed in the N2 method. This idealization is appropriate for bridges with roller supports at the abutments (dashed line in Fig. 3.7b). In bridges pinned at the abutments the bilinear idealization is more adequate (dashed line in Fig. 3.7a), since the corresponding capacity curve can have very steep hardening slope.

Idealization of the capacity curve is one of the basic steps of the N2 method, since it significantly influences the stiffness of the equivalent SDOF model and the value of the maximum displacement. When this stiffness is not adequately calculated the actual and estimated maximum displacement could be significantly different (Isaković and Fischinger 2006).

The capacity curve is usually idealized using the equal energy principle of idealized and actual curve. Since the energy depends on the reached maximum displacement (target displacement), which is not known at the moment of the idealization, iterations are necessary. In the majority of cases usually only one iteration is needed.

In the EC8/1 these iterations are optional. The informative annex H of the EC8/2 proposes to estimate the maximum displacement (target displacement) using the results of the linear multimode spectrum analysis. This solution is very convenient at the first glance. However, it demands a proper estimation of the effective stiffness of the columns corresponding to a certain seismic intensity. In bridges that are supported by columns of very different stiffness and strength this procedure requires iterations as well, since in such bridges it is quite difficult to estimate the effective stiffness adequately in advance. Consequently, the calculation procedure could be more time consuming than that proposed by Isaković et al.

3.4.1.3 Concluding Remarks for the N2 Method

The N2 method is the standard pushover method included into the Eurocode standards. It makes the nonlinear seismic analysis of bridges (structures) straightforward, since the dynamic analysis can be performed on the simplified model using the nonlinear response spectra.

There are several variants of the method proposed. Here, the variant, which is according to the authors' opinion the most appropriate for bridges, is presented. Compared to the basic variant, it includes certain modifications that take into account specifics of the bridge response. Some of these modifications are proposed in Eurocode 8 – Part 2 and some of them are proposed by Isaković and Fischinger (2006).

Specifics in the application of the N2 method for the analysis of bridges are related to: (1) the choice of the point, where the displacements of the MDOF model of the bridge are monitored during the static nonlinear analysis, (2) the choice of the distribution of the lateral forces along the bridge, used in the static pushover analysis, (3) idealization of the capacity curve.

1. In general, the point where the displacements are monitored to construct the pushover curve is the position of the maximum displacement of the superstructure.
2. In bridges, which are pinned at the abutments, uniform distribution, parabolic distribution or distribution of inertial forces proportional to the most important mode in the elastic range can be used. In bridges with roller supports at the abutments, the uniform distribution or distribution proportional to the (instantaneous) most important mode shape is appropriate. It is recommended to perform the analysis using two different distributions and to consider the envelope of the related response.
3. Due to the specific shape of the force-displacement relationship, the bilinear idealization of the capacity curve is appropriate for bridges that are pinned at the abutments. For bridges with roller supports at the abutments the elastoplastic idealization can be used. In general, an iterative approach is needed when the capacity curve is idealized, since this idealization depends on the target displacement (maximum reached displacement at certain seismic intensity level) which is not known in advance.

3.4.2 *Single-mode Pushover Analysis Procedure with Adaptive Load or Displacement Patterns*

As indicated above, probably the most critical problem of pushover analysis in bridge applications is the choice of the *monitoring point* to identify the horizontal axis of the conventional pushover curve. This problem actually results from the adoption of the invariant load patterns in pushover analysis. An adaptive version of the single-mode pushover method may be particularly attractive in bridge applications, in which plotting of the conventional pushover curve is completely avoided and the procedure leads directly to the *modal capacity diagram* for spectral demand estimation (Aydınoğlu 2005; Aydınoğlu and Önem 2010).

In the case of adaptive approach, first-mode equivalent seismic load increment can be written for the (i)'th pushover step as

$$\Delta \mathbf{f}_1^{(i)} = \overline{\mathbf{m}}_1^{(i)} \Delta a_1^{(i)}; \quad \overline{\mathbf{m}}_1^{(i)} = \mathbf{M}\boldsymbol{\Phi}_1^{(i)}\Gamma_{x1}^{(i)} \tag{3.11}$$

where $\mathbf{M}$ is the mass matrix, $\boldsymbol{\Phi}_1^{(i)}$ represents the instantaneous mode shape vector and $\Gamma_{x1}^{(i)}$ denotes the participation factor for the first (predominant) mode at the (i)'th step for an earthquake in x direction. In Eq. 3.11 $\overline{\mathbf{m}}_1^{(i)}$ represents the vector of *participating modal masses* and $\Delta a_1^{(i)}$ denotes the instantaneous *modal pseudo-acceleration increment* effective in the first mode. Superscript (i) indicates that instantaneous first-mode shape corresponding to the current configuration of the structural system is considered following the formation of the last plastic hinge at the end of the previous pushover step.

On the other hand, a fully compatible modal expression can be written for the increment of displacement vector as

$$\Delta \mathbf{u}_1^{(i)} = \overline{\mathbf{u}}_1^{(i)} \Delta d_1^{(i)}; \quad \overline{\mathbf{u}}_1^{(i)} = \boldsymbol{\Phi}_1^{(i)}\Gamma_{x1}^{(i)} \tag{3.12}$$

where $\Delta d_1^{(i)}$ denotes the instantaneous *modal displacement increment* effective in the first mode. Note that instantaneous modal pseudo-acceleration and displacement increments are interrelated as

$$\Delta a_1^{(i)} = (\omega_1^{(i)})^2 \Delta d_1^{(i)} \tag{3.13}$$

where $\omega_1^{(i)}$ refers to the instantaneous first-mode natural frequency.

Since $\Delta \mathbf{f}_1^{(i)}$ and $\Delta \mathbf{u}_1^{(i)}$ given in Eqs. 3.11 and 3.12, respectively, are both based on the same instantaneous modal quantities, a one-to-one correspondence always exists between them. Accordingly, adaptive implementation of the single-mode pushover analysis can be based on either a monotonic increase of displacements or equivalent seismic loads, leading to the displacement-controlled or load-controlled adaptive analyses (Aydınoğlu and Önem 2010). In the former, instantaneous modal displacement increment $\Delta d_1^{(i)}$ is obtained at the end of each pushover step, whereas instantaneous modal pseudo-acceleration increment $\Delta a_1^{(i)}$ would be the corresponding

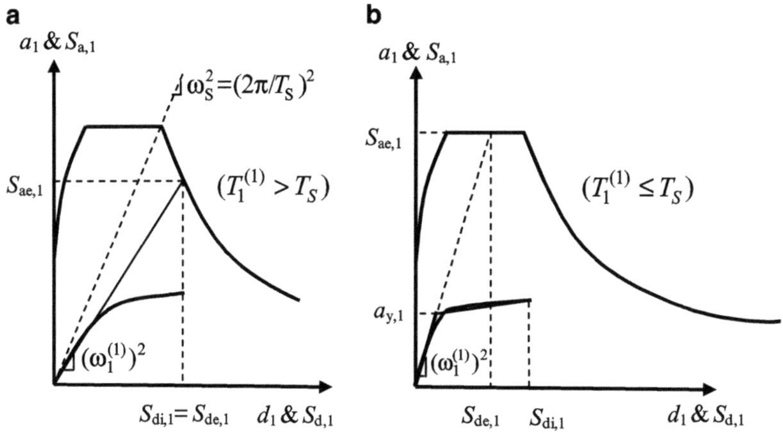

Fig. 3.8 Estimating modal displacement demand

output of the load-controlled adaptive analysis. In each case, the other modal quantity is obtained from Eq. 3.13. Cumulative quantities are then obtained as

$$d_1^{(i)} = d_1^{(i-1)} + \Delta d_1^{(i)} \tag{3.14}$$

$$a_1^{(i)} = a_1^{(i-1)} + \Delta a_1^{(i)} \tag{3.15}$$

Note that modal capacity diagram of the equivalent SDOF system is directly obtained in terms of a_1 versus d_1 without actually plotting the pushover curve in terms of total seismic force versus monitoring point displacement of the bridge. Thus, the problem of identifying the monitoring point is completely eliminated. Modal capacity diagram is then superimposed with the elastic acceleration-displacement response spectrum (ADRS) in terms of spectral displacement and spectral pseudo-acceleration as shown in Fig. 3.8 to estimate the modal displacement demand with a similar method described above in Sect. 3.4.1. Figure 3.8 is excerpted from the Turkish code for the seismic design of railway bridges (Ministry of Transportation 2008).

The displacement controlled single-mode adaptive analysis will also be presented in Sect. 3.5.2.5 as a special case of multi-mode adaptive pushover analysis procedure IRSA Method, where details of plastic hinge identification at each pushover step are explained.

3.5 Multi-mode Pushover Analysis Procedures

It is clear that single-mode pushover analysis can be reliably applied to only two-dimensional response of regular bridges (or low-rise building structures regular in plan), where seismic response is essentially governed by the fundamental mode.

Application of single-mode pushover to irregular bridges involving three-dimensional response (or to high-rise buildings or any building irregular in plan) could lead to incorrect, unreliable results. Therefore, a number of improved pushover analysis procedures have been offered in recent years in an attempt to take higher mode effects into account.

Regarding the multi-mode pushover analysis procedures, the approximate extension of the Response Spectrum Analysis (RSA) for inelastic analysis is developed in three different routes, as explained in the following sections.

3.5.1 Multi-mode Procedure Based on Independent Modal Pushover Analyses with Invariant Load Patterns: The MPA (Modal Pushover Analysis) Method

The Modal Pushover Analysis (MPA) Method is introduced in a systematic way by Chopra and Goel (2002) for building-type structures, although the basic idea behind the procedure was in fact proposed in earlier studies (Paret et al. 1996; Sasaki et al. 1998). The basic steps of MPA can be summarized as follows:

(i) Perform pushover analyses and plot pushover curves independently for each individual mode considered with invariant lateral load patterns associated with the linear (initial) mode shape;
(ii) Convert pushover curves in each mode to capacity diagrams of corresponding equivalent single-degree-of-freedom (SDOF) systems using modal conversion parameters based on the same linear (initial) mode shape;
(iii) Calculate inelastic seismic demands of equivalent SDOF systems in each mode independently through response history analysis or inelastic response spectrum;
(iv) Calculate mode contributions of inelastic modal response quantities of interest in each mode independently;
(v) Combine mode contributions with well-known modal combination rules to approximately obtain inelastic seismic demand quantities.

MPA was recently applied to bridges by Paraskeva et al. (2006), Isakovic and Fischinger (2006, 2011), Isakovic et al. (2008), Paraskeva and Kappos (2008, 2010). The procedure by Paraskeva and Kappos is described in some detail in the following.

3.5.1.1 Problems Encountered in Applying the MPA Procedure to Bridges

A key issue in applying the MPA procedure to bridges is the selection of an appropriate control point to monitor the displacement demand and also to plot the pushover curve for each mode. Natural choices for the monitoring point in a bridge are the deck mass centre (CEN 2004), or the top of the nearest pier, if the displacement of the two is practically the same (monolithic connection). Another proposal for the monitoring point of the bridge is the point of the deck where the (current) displacement is maximum (Fischinger et al. 2004; Isakovic and Fischinger 2006). By analogy to buildings (Chopra 2001) it can also be selected as the point of the deck that corresponds to the location along the longitudinal axis of the bridge of an equivalent SDOF system, defined by the location of the resultant of the pertinent modal load pattern (see Sect. 3.5.1.2). Several choices of monitoring point are acceptable as long as the derived pushover curve has a reasonable (bilinear) shape.

Another issue is the number of modes that should be considered in the case of bridges. Judging the required number of modes is far from straightforward in the case of bridges. Capturing all modes whose effective masses add up to 90% of the total mass of a bridge structure might need considering up to a very high number of modes. Work carried out by Paraskeva et al. (2006) has shown that there is little merit in adding modes whose participation factor is very low (say less than 1%), and less rigid rules than the 90% one (calibrated only for buildings) could be adopted. Further critical issues in MPA that influence the accuracy of the method are the way any pushover curve is bilinearized before being transformed into a capacity curve, and the method to be used for defining the earthquake demand for each mode. In the study of Paraskeva et al. (2006), the capacity and demand spectra method (CDSM) based on the use of inelastic spectra is recommended; details are given in the next section.

In developing the MPA procedure for bridges, wherein higher modes usually play a critical role, it was found that both the target displacement and the bridge response quantities were dependent on the selected monitoring point. To overcome this problem, which is associated with the inelastic range of modal pushover curves for higher modes, an improved MPA was proposed (Paraskeva and Kappos 2008, 2010), involving an additional step compared to the initial one, the key idea being that the deformed shape of the structure subjected to the considered earthquake level (to which it may respond inelastically) is used in lieu of the elastic mode shape; this improved version of the method is presented in the next section. To investigate the applicability of the improved MPA procedure for bridges, a number of actual bridge structures were studied, some of which are reported in Sect. 4 of this book. Comparative evaluation of the calculated response of different bridges illustrates the applicability and potential of the improved modal pushover method for bridges, and quantifies its relative accuracy compared to that obtained through other inelastic methods, especially for complex and irregular bridges.

3.5.1.2 Description of the Procedure Proposed for Bridges

The steps of the modified MPA procedure (including those that are the same as in the Chopra and Goel method) are summarized in the following.

Step 1: Compute the natural periods, T_n, and mode shapes, ϕ_n, for linearly elastic vibration of the structure.

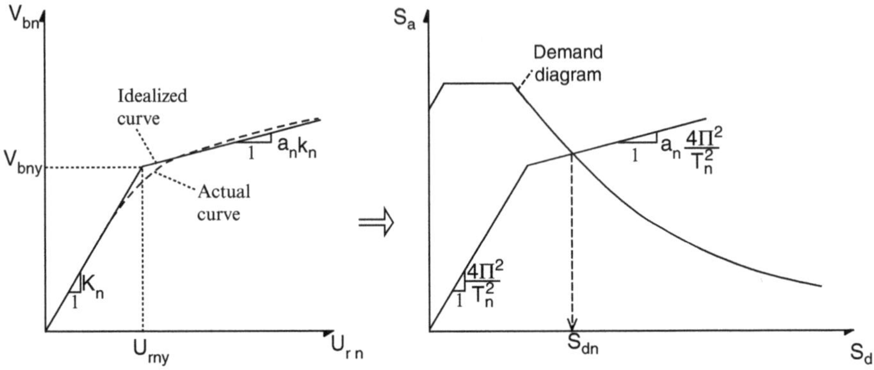

Fig. 3.9 Idealized pushover curve of the *n*th mode of the MDOF system, and corresponding capacity curve for the *n*th mode of the equivalent inelastic SDOF system

Step 2: Carry out separate pushover analyses for force distribution $\mathbf{s}_n^* = \mathbf{m} \cdot \phi_n$, for each significant mode of the bridge and construct the base shear vs. displacement of the 'control' or 'monitoring' point (V_{bn} vs. u_{cn}) pushover curve for each mode. It is worth stressing that $\mathbf{s}_n^*$ are loading *patterns*, hence the relative significance of each mode is not accounted for at this stage; this will be done at Step 5, through the target displacement for each modal pushover analysis. Gravity loads are applied before each MPA and P-Δ effects are included, if significant (e.g. in bridges with tall piers). It is noted that the value of the lateral deck displacement due to gravity loads, u_{rg}, is negligible for a bridge with nearly symmetrically distributed gravity loading.

Step 3: The pushover curve must be idealized as a bilinear curve so that a yield point and ductility factor can be defined and then be used to appropriately reduce the elastic response spectra representing the seismic action considered for assessment. This idealization can be done in a number of ways, some more involved than others; it is suggested to do this once (as opposed, for instance, to the ATC (1996) procedure) using the full pushover curve (i.e. analysis up to 'failure' of the structure, defined by a drop in peak strength of about 20%) and the equal energy absorption rule (equal areas under the actual and the bilinear curve). It is noted that the remaining steps of the proposed methodology can be applied even if a different method for producing a bilinear curve is used.

Step 4: Several procedures are available (ATC 1996; Chopra and Goel 2002; CEN 2005; FEMA 2005) for defining the earthquake displacement demand associated with each of the pushover curves derived in Step 3. Here the procedure used for estimating the displacement demand at the monitoring point is the version of the capacity spectrum method (CSM) based on inelastic demand spectra (Fajfar 1999; Chopra and Goel 1999); hence Step 4 consists in converting the idealized $V_{bn} - u_{cn}$ pushover curve of the multi-degree-of freedom (MDOF) system to a 'capacity diagram' (Fig. 3.9). The base shear forces and the corresponding displacements in each pushover curve are converted to spectral acceleration (S_a) and spectral displacements (S_d) respectively, of an equivalent single-degree-of-freedom (SDOF) system, using the relationships (ATC 1996; Chopra and Goel 2002):

$$S_a = \frac{V_{bn}}{M_n} \tag{3.16}$$

$$S_d = \frac{u_{cn}}{\Gamma_n \cdot \phi_{cn}} \tag{3.17}$$

wherein ϕ_{cn} is the value of ϕ_n at the control (or 'monitoring') point, $M_n^* = L_n \cdot \Gamma_n$ is the effective modal mass, with $L_n = \boldsymbol{\phi}_n{}^T \mathbf{m} \cdot \mathbf{1}$, $\Gamma_n = L_n/M_n$ is a mass participation factor, and $M_n = \boldsymbol{\phi}_n{}^T \mathbf{m} \cdot \boldsymbol{\phi}_n$ is the generalized mass, for the nth natural mode. For *inelastic* behaviour, the procedure used in the present study for estimating the displacement demand at the monitoring point is based on the use of inelastic spectra; this is equally simple, more consistent, and more accurate than the 'standard' capacity spectrum method adopted by ATC (1996) that is based on reducing the elastic spectra with ductility–dependent damping factors, as shown in a number of studies (Fajfar 1999; Chopra and Goel 1999; Kappos and Petranis 2001).

Step 5: Since the displacement demand calculated in Step 4 (for each mode) refers to SDOF systems with periods equal to those of the corresponding modes, the next step is to correlate these displacements to those of the actual bridge. Hence, Step 5 consists in converting the displacement demand of the nth mode inelastic SDOF system to the peak displacement of the monitoring point, u_{cn} of the bridge, using Eq. 3.17. The selection of this point is a critical issue for MPA of bridges and is discussed in Paraskeva et al. (2006). Several choices of monitoring point are acceptable as long as the derived pushover curve has a reasonable shape, but they do not lead to equally good results as far as the final response quantities are concerned (Lupoi et al. 2007).

Step 6: In this step, a correction is made of the displacement of the monitoring point of the bridge, which was calculated at the previous step. The correction is necessary only for cases that significant inelasticity develops in the structure. If the structure remains elastic or close to the yield point, the MPA procedure suggested by Paraskeva et al. (2006) is used to estimate seismic demands for the bridge. The response displacements of the structure are evaluated by extracting from the database of the individual pushover analyses the values of the desired responses at which the displacement at the control point is equal to u_{cn} (see Eq. 3.17). These displacements are then applied to derive a new vector ϕ_n', which is the deformed shape (affected by inelastic effects) of the bridge subjected to the given modal load pattern. The target displacement at the monitoring point for each pushover analysis is calculated again with the use of ϕ_n', according to Eq. 3.18

$$u_{cn}' = \Gamma_n' \cdot \phi_n' \cdot S_{dn}' \tag{3.18}$$

wherein Γ_n' is Γ_n recalculated using ϕ_n', and S'_{dn} is the displacement of the equivalent SDOF system (which generally differs from S_{dn}).

Step 7: The response quantities of interest (displacements, plastic hinge rotations, forces in the piers) are evaluated by extracting from the database of the individual pushover analyses the values of the desired responses r_n, due to the combined effects of gravity and lateral loads for the analysis step at which the displacement at the control point is equal to u_{cn} (see Eq. 3.3).

Step 8: Steps *3* to *7* are repeated for as many modes as required for sufficient accuracy. As noted in the previous section, there is little merit in adding modes whose participation factor is very low (say less than 1%), and application of the method to a number of bridges shows that it is not necessary to assure that the considered modes contribute to 90% of the total mass.

Step 9: The total value for any desired response quantity (and each level of earthquake intensity considered) can be determined by combining the peak 'modal' responses, r_{no} using an appropriate modal combination rule, e.g. the SRSS combination rule, or the CQC rule. This simple procedure was used here for both displacements and plastic hinge rotations, which were the main quantities used for assessing the bridges analysed (whose response to service gravity loading was, of course, elastic). If member (e.g. pier) forces have to be determined accurately (in an inelastic procedure this is equivalent to determining the percentage by which yield strength of the members is exceeded), a more involved procedure of combining modal responses should be used. Such a procedure was suggested by Goel and Chopra (2004) for buildings, consisting essentially in correcting the bending moments at member ends (whenever yield values were exceeded) on the basis of the relevant moment – rotation diagram and the value of the calculated plastic hinge rotation. This procedure, which blends well with the capabilities of currently available software, has also been used in the case studies presented in Chap. 4.

3.5.1.3 Derivation of Pushover Curves Using MPA and Comparison with Dynamic Analysis

In seismic assessment of structures the pushover (or, better, the resistance) curve is a key tool, in the sense that it provides a good description of both the strength and the available ductility of the structure, both combined into a single diagram. Clearly, when the loading pattern used for producing such a curve is inadequate (i.e. when a single mode pattern is not sufficient for identifying the salient features of the response) one cannot expect to obtain a reliable pushover curve. In view of these remarks the concept of multimodal pushover curve (as part of the MPA procedure) in terms of base shear vs. deck displacement is introduced herein for the case of bridges. The concept is also valid for buildings, and indeed, to the authors' best knowledge this is the first time that such a 'composite' pushover curve is introduced for any type of structure. Han and Chopra (2006) have used the MPA procedure for producing Incremental Dynamic Analysis (IDA) curves, i.e. plots of an earthquake intensity measure (elastic S_a corresponding to the fundamental period T_1 of the building) vs. an engineering demand parameter (roof drift ratio).

These curves are different from the multimodal pushover curves presented herein; the main use of IDA curves (Vamvatsikos and Cornell 2002) is in estimating demands for a given earthquake intensity, whereas multi-modal pushover curves are primarily intended for estimating the capacity of a structure in terms of strength and ductility.

A multimodal pushover curve cannot be derived from a simple combination of the individual curves derived for each modal pattern; it is essential that the earthquake demand for each stage of the bridge's response is properly accounted for, so that forces and displacements from each modal curve can indeed be combined (in a statistical sense, of course). Hence, for the estimation of multimodal pushover curves, a significant number of analyses for different levels of earthquake intensity has to be carried out, i.e. the base shear, as well as the deck displacement at the selected control point, are first calculated for each mode independently for each level of earthquake intensity (see Step 4 of the procedure described in the previous section). The value of the total displacement can then be determined by combining the peak 'modal' responses, using the SRSS combination rule. Regarding base shear, as well as forces in the structural members, these should not be determined by SRSS combination, as this would significantly overestimate them in most cases; instead, the procedure suggested by Goel and Chopra (2004) can be used, consisting essentially in correcting the bending moments at plastic hinge locations on the basis of the relevant moment-rotation diagram and the value of the plastic hinge rotation calculated from the SRSS combination. The procedure for the derivation of a multimodal pushover curve is summarized in the following.

- Response quantities of interest (displacements of the selected monitoring point, plastic hinge rotations, shear forces in the piers) are calculated for each mode, using Steps 1–8 of the methodology introduced in the previous section; this is repeated for as many modes as required for sufficient accuracy (see Step 4 of the proposed MPA procedure).
- The value of the deck displacement of the selected control point is determined, for each level of earthquake intensity considered, by combining the modal displacements of the control point u_{cn} (or u'_{cn} if Eq. 3.18 is used), using a statistical combination rule. If the structure remains elastic for the considered earthquake intensity level, then the value of the base shear of the structure is determined using the same procedure. However, if the structure enters the inelastic range for the considered earthquake intensity level, a more involved procedure of combining modal shear forces in the piers is used. First, the total value of the plastic hinge rotation θ_{pj} at each pier end is estimated as the SRSS combination of the modal values θ_{pjn}. The corresponding bending moments in the piers are estimated through the relevant moment-rotation diagram at the value of the plastic hinge rotation calculated from the SRSS combination. Then, shears in the piers are calculated using the corrected bending moments, and the base shear is calculated as the sum of the pier shears (in the direction under consideration).
- The above procedure is repeated for as many earthquake intensity levels as required for drawing a representative multimodal pushover curve. Derivation of

a 'complete' multimodal pushover curve up to 'failure' of the structure (defined by a drop in peak strength of about 20%) is recommended for seismic assessment of the bridge.

- The multimodal pushover curve is derived by plotting the total value of deck displacement of the selected control point against the corresponding value of the base shear of the structure, for each earthquake intensity level.

Validation of the aforementioned multimodal pushover curve can be made by comparing it to 'dynamic' pushover curves, derived from NRHA, which should not be confused with an IDA curve. A dynamic pushover curve can be derived by extracting from the 'database' of time-history analysis the values of the desired response quantities, i.e. base shear of the structure and displacements of the selected control point. The deck displacement of the control point is the average of the maximum displacements of the selected point recorded in the structure during the response history analyses for a number of accelerograms. Three different combinations of base shear and maximum displacement of the monitoring point were used by Paraskeva & Kappos (2008) in order to derive dynamic pushover curves:

- Maximum displacement (u_{max}) at the control point vs. simultaneous base shear of the bridge ($V_b(t)$).
- Maximum displacement (u_{max}) at the control point vs. the base shear ($V_b(t-\Delta t)$) recorded at the previous step of that when u_{max} was recorded; or the base shear recorded after the step of the maximum displacement at the control point ($V_b(t+\Delta t)$). In a previous study concerning asymmetric buildings (Penelis and Kappos 2005) this 'time-window' approach was found to give the most meaningful dynamic pushover curves.
- Maximum displacement (u_{max}) at the control point vs. maximum base shear of the bridge (V_{bmax}). This is considered only as an upper bound, since, obviously, these response quantities are not simultaneous.

It is noted that, in order to compare a 'standard' pushover curve (derived by SPA), to a multimodal pushover curve (derived by the MPA procedure or the improved MPA procedure wherever necessary), or to a dynamic pushover curve (derived by NRHA), the selected monitoring point of deck displacements has to be common for all curves.

As will be shown by the case studies presented in Chap. 4, the multimodal pushover curve reasonably matches the dynamic pushover curves derived from the more rigorous NRHA.

3.5.2 Simultaneous Multi-mode Procedure with Modal Adaptive Displacement Patterns: The IRSA (Incremental Response Spectrum Analysis) Method

A multi-mode adaptive pushover procedure called Incremental Response Spectrum Analysis (IRSA) Method was developed by Aydınoğlu (2003, 2004), in which modal pushovers are implemented *simultaneously* by imposing instantaneous displacement increments at each pushover step. In principle, modal displacements are scaled in IRSA with respect to *inelastic spectral displacements* that are associated with the *instantaneous* configuration of the structure. In practice, modal scaling is favourably achieved through *elastic spectral displacements* by taking advantage of the well-known *equal displacement rule.*

Applications of IRSA to bridges can be found in Aydınoğlu (2004), Aydınoğlu and Önem (2009), Isakovic and Fischinger (2006, 2011), Isakovic et al. (2008). The main aspects of the IRSA Method are described in the following.

3.5.2.1 Piecewise Linear Modal Representation of Nonlinear Response

Incremental Response Spectrum Analysis (IRSA) is a multi-mode pushover procedure, in which the incremental response is assumed *piecewise linear* at each pushover step in-between the formation of two consecutive plastic hinges.

In a piecewise linear response in a typical n'th mode, the *instantaneous* increments of the equivalent seismic load vector, $\Delta\mathbf{f}_n^{(i)}$, and the corresponding displacement vector, $\Delta\mathbf{u}_n^{(i)}$, at the (i)'th incremental pushover step can be expressed as,

$$\Delta\mathbf{f}_n^{(i)}=\mathbf{M}\boldsymbol{\Phi}_n^{(i)}\,\Gamma_{xn}^{(i)}\Delta a_n^{(i)} \tag{3.19}$$

$$\Delta\mathbf{u}_n^{(i)}=\boldsymbol{\Phi}_n^{(i)}\Gamma_{xn}^{(i)}\Delta d_n^{(i)} \tag{3.20}$$

where $\mathbf{M}$ represents the mass matrix, $\boldsymbol{\Phi}_n^{(i)}$ is the n'th instantaneous mode shape vector at the (i)'th piecewise linear incremental pushover step and $\Gamma_{xn}^{(i)}$ refers to the corresponding participation factor for an x-direction earthquake. Adaptive (instantaneous) n'th mode shape vector satisfies the following free-vibration equation:

$$(\mathbf{K}^{(i)}-\mathbf{K}_G^{(i)})\boldsymbol{\Phi}_n^{(i)}=(\omega_n^{(i)})^2\mathbf{M}\boldsymbol{\Phi}_n^{(i)} \tag{3.21}$$

where $\mathbf{K}^{(i)}$ and $\mathbf{K}_G^{(i)}$ represent, respectively, the instantaneous first-order stiffness matrix and geometric stiffness matrix, the latter of which is included to consider the second-order (P-Delta) effects. $\omega_n^{(i)}$ is the circular natural frequency of the structure based on hinge distribution at the (i)'th piecewise linear incremental pushover step.

$\Delta d_n^{(i)}$ and $\Delta a_n^{(i)}$ in Eq. 3.19 and Eq. 3.20 represent n'th *modal displacement and modal pseudo-acceleration increments*, respectively, at the (i)'th pushover step, which are interrelated as follows:

$$\Delta a_n^{(i)}=(\omega_n^{(i)})^2\Delta d_n^{(i)} \tag{3.22}$$

Cumulative values of modal displacement and pseudo-acceleration are calculated at each step by adding the incremental values to the cumulative values obtained at the end of the previous step:

$$d_{\mathrm{n}}^{(\mathrm{i})}=d_{\mathrm{n}}^{(\mathrm{i}-1)}+\Delta d_{\mathrm{n}}^{(\mathrm{i})} \tag{3.23}$$

$$a_{\mathrm{n}}^{(\mathrm{i})}=a_{\mathrm{n}}^{(\mathrm{i}-1)}+\Delta a_{\mathrm{n}}^{(\mathrm{i})} \tag{3.24}$$

Note that, similar to single-mode adaptive pushover analysis described in Sect. 3.4.2, a one-to-one correspondence exists between the incremental seismic loads and the resulting displacements in each mode given by Eqs. 3.19 and 3.20, respectively. In other words, seismic load increments given by Eq. 3.19 always result in displacement increments represented by Eq. 3.20. It is for this reason that force-controlled or displacement-controlled pushovers are equally applicable in adaptive incremental schemes.

The aim of the piecewise linear modal pushover analysis is the determination of incremental and cumulative values of modal displacements and modal pseudo-accelerations in each mode, which eventually define *modal capacity diagrams* (See Fig. 3.10). Peak values of modal displacements may be obtained from the following incremental modal equation of motion

$$\Delta\ddot{d}_{\mathrm{n}}^{(\mathrm{i})}+2\xi_{\mathrm{n}}^{(\mathrm{i})}\omega_{\mathrm{n}}^{(\mathrm{i})}\Delta\dot{d}_{\mathrm{n}}^{(\mathrm{i})}+\Delta a_{\mathrm{n}}^{(\mathrm{i})}=-\,\Delta\ddot{u}_{\mathrm{x}}^{\mathrm{g}(\mathrm{i})} \tag{3.25}$$

where $\Delta\ddot{u}_{\mathrm{x}}^{\mathrm{g}(\mathrm{i})}=\ddot{u}_{\mathrm{x}}^{\mathrm{g}(\mathrm{i})}-\ddot{u}_{\mathrm{x}}^{\mathrm{g}(\mathrm{i}-1)}$ is the ground acceleration increment and $\xi_{\mathrm{n}}^{(\mathrm{i})}$ refers to modal damping ratio.

3.5.2.2 Modal Scaling

In order to define modal MDOF response, modal displacement increments $\Delta d_{\mathrm{n}}^{(\mathrm{i})}$ or modal pseudo-acceleration increments $\Delta a_{\mathrm{n}}^{(\mathrm{i})}$ have to be determined in all modes at each pushover step, depending on whether displacement- or force-controlled pushover is applied.

Since just a single plastic hinge forms and therefore only one yield condition is applicable at the end of each piecewise linear step, a reasonable assumption needs to be made for the relative values of modal displacement or modal pseudo-acceleration increments, so that the number of unknowns can be reduced to one. This is called *modal scaling*, which is the most critical assumption to be made in all multi-mode pushover procedures with the sole exception of Modal Pushover Analysis–MPA (Chopra and Goel 2002; Paraskeva et al. 2006) where modal coupling is completely disregarded in the formation of plastic hinges.

Displacement-controlled pushover is the preferred approach in IRSA where modal pushovers are implemented simultaneously by imposing instantaneous displacement increments of MDOF system at each pushover step according to Eq. 3.20.

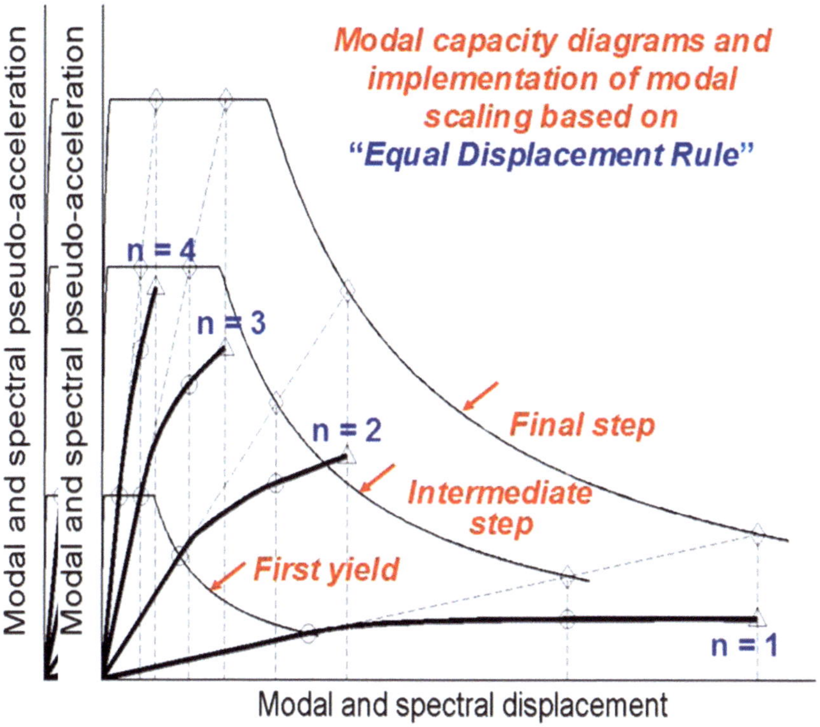

Fig. 3.10 Modal capacity diagrams and scaling of modal displacements through monotonic scaling of response spectrum

In principle, modal displacements are scaled in IRSA with respect to *inelastic spectral displacements*, $S_{\mathrm{din}}^{(i)}$, associated with the *instantaneous* configuration of the structure (Aydınoğlu 2003). This is the main difference of IRSA from other multi-mode pushover procedures where modal scaling is based on *instantaneous elastic* spectral pseudo-accelerations or displacements (see Sect. 3.5.3 below).

Scaling Based on Equal Displacement Rule

In practice, modal scaling based on *inelastic spectral displacements* can be easily achieved by taking advantage of the *equal displacement rule*. Assuming that seismic input is defined via *smoothed elastic response spectrum*, according to this simple and well-known rule, *peak displacement* of an inelastic SDOF system and that of the corresponding elastic system are assumed practically equal to each other provided that the effective initial period is longer than the *characteristic period* of the elastic response spectrum. The characteristic period is approximately defined as the transition period from the constant acceleration segment to the constant velocity segment of the spectrum. For periods shorter than the characteristic period, elastic spectral displacement is amplified using a displacement modification factor, i.e., C_1 coefficient given in FEMA 356 (ASCE 2000). However such a situation is seldom encountered in bridges involving multi-mode response. In such structures, effective initial periods of the first few modes are likely to be longer than the characteristic period and therefore those modes automatically qualify for the equal displacement rule. On the other hand, effective post-yield slopes of the modal capacity diagrams get steeper and steeper in higher modes with gradually diminishing inelastic behaviour (Fig. 3.10). Thus it can be comfortably assumed that inelastic spectral displacement response in higher modes would not be different from the corresponding spectral elastic response. Hence, smoothed elastic response spectrum may be used in its entirety for scaling modal displacements without any modification.

Note that in practice cracked section stiffnesses are used in reinforced concrete systems throughout the pushover analysis and thus elastic periods calculated at the first piecewise-linear pushover step (i = 1) may be considered in lieu of the initial periods generally obtained in practice from bi-linearization of modal capacity diagrams.

In line with the *equal displacement rule*, scaling procedure applicable to n'th mode increment of modal displacement at the (i)'th pushover step is simply expressed as

$$\Delta d_{\mathrm{n}}^{(\mathrm{i})}=\Delta\tilde{F}^{(\mathrm{i})}S_{\mathrm{den}}^{(1)} \tag{3.26}$$

where $\Delta\tilde{F}^{(\mathrm{i})}$is an *incremental scale factor*, which is applicable to all modes at the (i)'th pushover step. $S_{\mathrm{den}}^{(1)}$ represents the *initial elastic spectral displacement* defined at the first step, which is taken equal to the *inelastic spectral displacement* associated with the *instantaneous* configuration of the structure at any pushover step. Cumulative modal displacement at the end of the same pushover step can then be written as

$$d_{\mathrm{n}}^{(\mathrm{i})}=\tilde{F}^{(\mathrm{i})}S_{\mathrm{den}}^{(1)} \tag{3.27}$$

in which $\tilde{F}^{(\mathrm{i})}$represents the *cumulative scale factor* with a maximum value of unity:

$$\tilde{F}^{(\mathrm{i})}=\tilde{F}^{(\mathrm{i}-1)}+\Delta\tilde{F}^{(\mathrm{i})}\leq 1 \tag{3.28}$$

Monotonic Scaling of Response Spectrum

Note that modal scaling expressions given above correspond to a monotonic increase of the elastic response spectrum progressively at each step with a cumulative scale factor starting from zero until unity. Physically speaking, the structure is being pushed such that at every pushover step modal displacements of all modes are increased by increasing elastic spectral displacements defined at the first step (i = 1) in the same proportion according to *equal displacement rule* until they simultaneously reach the target *spectral displacements* on the response spectrum. Shown in Fig. 3.10 are the scaled spectra representing the first yield, an intermediate pushover step ($\tilde{F}^{(i)} < 1$) and the final step ($\tilde{F}^{(i)} = 1$), which are plotted in ADRS (Acceleration-Displacement Response Spectrum) format and superimposed onto modal capacity diagrams. At this point, it is worth warning that equal displacement rule may not be valid at near-fault situations with forward directivity effect.

It needs to be stressed that since the above-mentioned monotonic spectrum scaling is based on the *equal displacement rule*, it is valid for spectral displacements only, not for the elastic spectral pseudo-accelerations. In other words, modal pseudo-accelerations have no relation at all with the elastic spectral accelerations, as seen in Fig. 3.10.

3.5.2.3 Multi-mode Pushover History Analysis: Simultaneous Modal Pushovers in All Modes with Modal Combination at Each Step

Substituting Eq. 3.26 into Eq. 3.20 leads to the following expression for the displacement vector increment in the n'th mode at the (i)'th pushover step:

$$\Delta\mathbf{u}_{\mathrm{n}}^{(\mathrm{i})} = \tilde{\mathbf{u}}_{\mathrm{n}}^{(\mathrm{i})} \Delta\tilde{F}^{(\mathrm{i})}; \quad \tilde{\mathbf{u}}_{\mathrm{n}}^{(\mathrm{i})} = \Phi_{\mathrm{n}}^{(\mathrm{i})} \Gamma_{\mathrm{xn}}^{(\mathrm{i})} S_{\mathrm{den}}^{(1)} \tag{3.29}$$

Now, piecewise linear multi-mode pushover history analysis can be performed at a given pushover step (i), by monotonically imposing displacement increments $\Delta\mathbf{u}_{\mathrm{n}}^{(\mathrm{i})}$ of MDOF system defined in Eq. 3.29 *simultaneously in all modes* considered. In this process, the increment of a *generic response quantity* of interest, such as the increment of an internal force, a displacement component, a story drift or the plastic rotation of a previously developed plastic hinge etc., may be calculated in each mode from the adaptive expression Eq. 3.29 as

$$\Delta r_{\mathrm{n}}^{(\mathrm{i})} = \tilde{r}_{\mathrm{n}}^{(\mathrm{i})} \Delta\tilde{F}^{(\mathrm{i})} \tag{3.30}$$

where $\tilde{r}_{\mathrm{n}}^{(\mathrm{i})}$ represents the generic response quantity to be obtained in each mode for $\Delta\tilde{F}^{(\mathrm{i})} = 1$, i.e., by imposing the displacement vector $\tilde{\mathbf{u}}_{\mathrm{n}}^{(\mathrm{i})}$ given in Eq. 3.29. Incremental scale factor $\Delta\tilde{F}^{(\mathrm{i})}$ is the only unknown at each pushover step corresponding to the formation of a new plastic hinge.

Modal Combination at Each Step

In the next stage, modal generic response quantity increments in each mode are combined by an appropriate modal combination rule:

$$\Delta r^{(\mathrm{i})} = \tilde{r}^{(\mathrm{i})} \Delta\tilde{F}^{(\mathrm{i})} \tag{3.31}$$

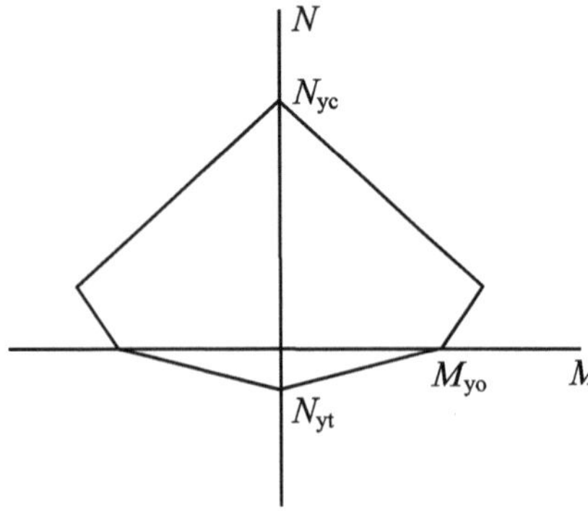

Fig. 3.11 A simplified representation of piecewise-linearized yield surface of a reinforced concrete section (number of linear segments may be increased for enhanced accuracy)

in which $\tilde{r}^{(i)}$ can be calculated with Complete Quadratic Combination (CQC) rule as

$$\tilde{r}^{(i)} = \sqrt{\sum_{m=1}^{N_m} \sum_{n=1}^{N_m} (\tilde{r}_m^{(i)} \rho_{mn}^{(i)} \tilde{r}_n^{(i)})} \tag{3.32}$$

where $\rho_{mn}^{(i)}$ is the cross-correlation coefficient of the CQC rule and N_m represents the number of modes considered in the analysis.

Thus, generic response quantity at the end of the (i)'th pushover step can be estimated as

$$r^{(i)} = r^{(i-1)} + \Delta r^{(i)} = r^{(i-1)} + \tilde{r}^{(i)} \Delta \tilde{F}^{(i)} \tag{3.33}$$

in which $r^{(i)}$and $r^{(i-1)}$are the *generic response quantities* to develop at the end of current and previous pushover steps, respectively. In the first pushover step (i = 1), response quantities due to gravity loading are considered as $r^{(0)}$.

Identification of Hinge Formation

In the next stage of IRSA, the generic expression given in Eq. 3.33 is specialized for the response quantities that define the coordinates of the *yield surfaces* of all potential plastic hinges, e.g., biaxial bending moments and axial forces in a general, three-dimensional response of a framed structure. As part of the piecewise linearization process of pushover analysis as well as to avoid iterative operations in hinge identification process, yield surfaces are appropriately linearized in a piecewise fashion, i.e., they are represented by finite number of lines or planes in two- and three-dimensional response models, respectively. As an example, planar yield surfaces (lines) of a reinforced concrete section (j) as shown in Fig. 3.11 where a typical line (s) can be expressed as

$$\alpha_{j,s} M_{jp} + \beta_{j,s} N_{jp} = 1 \tag{3.34}$$

in which M_{jp} and N_{jp} represents the yield bending moment and corresponding axial force, respectively, at section j while $\alpha_{j,s}$ and $\beta_{j,s}$ refer to the coefficients defining the yield line (s).

For the (i)'th pushover step, Eq. 3.33 is specialized for bending moment and axial force as

$$M_j^{(i)} = M_j^{(i-1)} + \tilde{M}_j^{(i)} \Delta\tilde{F}^{(i)}; \quad N_j^{(i)} = N_j^{(i-1)} + \tilde{N}_j^{(i)} \Delta\tilde{F}^{(i)} \tag{3.35}$$

which are then substituted into Eq. 3.34 with $M_j^{(i)} = M_{jp}$ and $N_j^{(i)} = N_{jp}$. Finally $\Delta\tilde{F}^{(i)}$ is extracted as,

$$(\Delta\tilde{F}^{(i)})_{j,s} = \frac{1 - \alpha_{j,s} M_j^{(i-1)} - \beta_{j,s} N_j^{(i-1)}}{\alpha_{j,s} \tilde{M}_j^{(i)} + \beta_{j,s} \tilde{N}_j^{(i)}} \tag{3.36}$$

The yield line (s) at section (j) that intersected with a minimum positive $(\Delta\tilde{F}^{(i)})_{j,s}$ among all yield lines of all potential plastic hinges identifies the new hinge formed at the end of the (i)'th pushover step.

Once $\Delta\tilde{F}^{(i)}$ is determined, any response quantity of interest developed at the end of that step can be obtained from the generic expression of Eq. 3.33. Modal displacement increment $\Delta d_n^{(i)}$ in any mode can be obtained from Eq. 3.26 and in turn modal pseudo-acceleration increment from Eq. 3.22, leading to the simultaneous estimation of respective cumulative quantities, i.e. the new coordinates of capacity diagrams in all modes, which can be obtained through Eqs. 3.23 and 3.24.

When the formation of the new hinge is identified, the current global stiffness matrix of the structure is locally modified such that only the element stiffness matrix affected by the new hinge is replaced with a new one for the next pushover step. Normality criterion is enforced in columns and wall type elements for the coupling of internal forces as well as the plastic deformation components of plastic hinges.

Estimation of Peak Quantities: Inelastic Seismic Demand

The above-described *pushover-history* process is repeated for all pushover steps until cumulative spectrum scale factor defined by Eq. 3.28 exceeds unity at the end of a given pushover step. When such a step is detected, which is indicated by superscript (p), incremental scale factor corresponding to this final pushover step is re-calculated from Eq. 3.28 as

$$\Delta\tilde{F}^{(p)} = 1 - \tilde{F}^{(p-1)} \tag{3.37}$$

Finally peak value of any response quantity of interest is obtained from the generic expression of Eq. 3.33 for $i = p$:

$$r^{(p)} = r^{(p-1)} + \Delta r^{(p)} = r^{(p-1)} + \tilde{r}^{(p)} \Delta\tilde{F}^{(p)} \tag{3.38}$$

3.5.2.4 Degeneration to Special Case: Single-mode Pushover Analysis Procedure with Adaptive Displacement Patterns

Single-mode pushover procedure based on adaptive displacement patterns, as described in Sect. 3.4.2, may be re-derived in this section as a special case of the IRSA Method. In this case all expressions for IRSA are written only for the first mode, i.e., n = 1 and modal scaling described in Sect. 3.5.2.2 as well as modal combination described in Sect. 3.5.2.3 becomes irrelevant. Identifying the formation of plastic hinges as well as the estimation of peak response quantities (inelastic seismic demand) are achieved by following exactly the same procedures given above for multi-mode IRSA.

A further special case of IRSA can be reached when the nonlinear system behaviour ceases, leading to a linear elastic response. In this case, single-mode or multi-mode IRSA degenerates to the standard *Response Spectrum Analysis (RSA)* procedure.

3.5.2.5 Treatment of P-Delta Effects in the IRSA Method

P-Delta effects are rigorously considered in IRSA through straightforward consideration of geometric stiffness matrix in each increment of the analysis procedure (see Eq. 3.21). Along the pushover-history process, accumulated plastic deformations may result in negative-definite second-order stiffness matrices, which in turn yield negative eigenvalues and hence negative post yield slopes in modal capacity diagrams of the lower modes (Aydınoğlu 2004). The corresponding mode shapes are representative of the post-buckling deformation state of the structure, which may significantly affect the distribution of internal forces and inelastic deformations of the structure.

Analysis of inelastic SDOF systems based on bilinear backbone curves with negative post-yield slopes indicates that such systems are susceptible to *dynamic instability* rather than having amplified displacements due to P-Delta effects. This has been first explored by Aydınoğlu and Fahjan (2003) and confirmed by Miranda and Akkar (2003). Consequently the use of P-Delta amplification coefficient (C_3) defined in FEMA 356 (ASCE 2000) has been cancelled (FEMA 2005, 2009; ASCE 2007). The dynamic instability is known to depend on the yield strength, initial stiffness, negative post-yield stiffness and the hysteretic model of SDOF oscillator as well as on the characteristics of the earthquake ground motion. Accordingly, practical guidelines have been proposed for minimum strength limits in terms of other parameters to avoid instability (Miranda and Akkar 2003; FEMA 2005, 2009; ASCE 2007). Further research is needed for the realistic cases of backbone curves resulting from modal capacity diagrams, which exhibit multiple post-yield slopes with both ascending and descending branches. For the time being, equal displacement rule is used in IRSA even when P-Delta effects are present as long as an imminent danger of dynamic instability is not expected according to the above-mentioned practical guidelines.

3.5.3 Multi-mode Procedures Based on Single-run Pushover Analysis with Modal Combined Adaptive Seismic Load or Displacement Patterns

In a number of adaptive pushover procedures that have been developed mainly for building structures in the last decade (e.g., Antoniou et al. 2002; Elnashai 2001; Antoniou and Pinho 2004a), equivalent seismic loads are calculated at each pushover step using mode shapes based on instantaneous (tangent) stiffness matrix and the corresponding *elastic spectral pseudo-accelerations*. Those modal seismic loads are then combined with a modal combination rule and normalized at each step to obtain an instantaneous *single load pattern* for a so-called *single-run pushover analysis* with an end product of *a single combined pushover curve*, which is supposed to represent the contributions of all modes considered. Alternatively, combined modal displacements are employed for the same purpose (Antoniou and Pinho 2004b; Pinho et al. 2007), which are defined at each pushover step based on instantaneous mode shapes and the corresponding *elastic spectral displacements*.

Whether elastic spectral accelerations or displacements associated with the instantaneous free vibration periods can be consistently used for an inelastic behaviour is a highly questionable assumption. But more importantly, an equivalent SDOF system leading to a *single capacity curve* is theoretically implausible to be deduced from a *single combined pushover curve*, since individual mode contributions to the seismic demand can no longer be de-aggregated. In fact, as it is clearly depicted in Fig. 3.10, modal capacity diagrams of the lower modes happen to be highly curved representing highly nonlinear behaviour with higher ductility capacities, whereas the diagrams of the higher modes tend to be straightened towards a linear behaviour. Recognizing this fact, majority of researchers used the above-described *single-run pushover analysis* procedure merely as a *multi-mode capacity estimation tool*, as mentioned earlier in Sect. 3.3.2. To verify the validity of capacity estimation, the structure is pushed to a *fixed target displacement* that is actually obtained earlier from a nonlinear response history analysis.

Yet few researchers maintained the idea that *single-run pushover analysis* procedure could also be used as a *demand estimation tool*. Accordingly, it is postulated that an SDOF capacity curve (sometimes called capacity spectrum) could be developed from the *combined pushover curve* with a modal conversion based on a combined deformed shape obtained from the single-run pushover analysis. In spite of the apparent theoretical inconsistency, some satisfactory results have been reported in seismic assessment of bridges (e.g., see Chap. 4). Thus, two adaptive *single-run pushover analysis* procedures are presented in this section, the first one being a load-controlled procedure and the second one a displacement-controlled procedure.

3.5.3.1 The Modal Adaptive Nonlinear Static Procedure (MANSP)

The MANSP is a load-controlled multi-mode pushover procedure, which was proposed and developed by Reinhorn (Bracci et al. 1997; Reinhorn 1997) and applied to bridges by DeRue (De Rue 1998). Comparison of the MANSP with some other pushover based methods can be found in Isakovic and Fischinger (2006).

The MANSP is an adaptive method since it takes into account the possibility that the mode shapes and the deflection line could change significantly when the structure is subjected to earthquakes of different intensity. The main steps of this method are summarized below:

1. Pushover analysis, which is used to calculate the capacity curve
2. Approximation of the capacity curve with an equivalent bi-linear model
3. Determination of the seismic demand using response spectra

Step 1: To obtain the pushover curve, pushover analysis with variable distribution of inertial forces should be performed. This distribution depends on the variable dynamic properties of the structure. The forces $q(u)_i$, which are applied to the structure are calculated combining the modal inertial forces as it is stated below:

$$q(u)_i = \boldsymbol{m}_i \, \Phi_{i1} \, \Gamma_1 S_{a_1} \left\| f_{ij} \gamma_j s_{aj} \right\|_p^{j=1,r} \tag{3.39}$$

where $q(u)_i$ is a force at certain location i of the structure, m_i is the corresponding mass, Φ_{i1} is the value of the first mode shape at location i, Γ_1 is participation factor of the first mode, S_{a1} is the spectral acceleration corresponding to the first mode, $f_{ij} = \Phi_{ij}/\Phi_{i1}$ is the ratio of the mode shape j and the first mode shape, $\gamma_j = \Gamma_j/\Gamma_1$ is the ratio of the participation factor of mode j and the first mode, $s_{aj} = S_{aj}/S_{a1}$ is the ratio of the spectral acceleration of the mode j and the first mode. In the inelastic range the modal characteristics Φ_j, γ_j, and S_{aj}, are continuously changing. Therefore, after each occurrence of a new plastic hinge, forces $q(u)_i$ are updated based on the new modal characteristics of the structure.
Different combinations of modal contributions can be used. When $p = 2$ is taken into account, the combination transforms to the well-known SRSS rule. Pushing the structure with forces $q(u)_i$ a pushover curve is obtained.

Step 2: In the next step the pushover curve is approximated by an equivalent bi-linear model (see Fig. 3.12). This model is determined considering the same post-elastic stiffness as in the original pushover curve and equal energy to failure.

Step 3: In the third step the pushover curve is transformed to spectral capacity curve. For the MDOF systems this transformation is performed as it is stated below:

(a) Normalized forces are determined as

$$Q*(u)_i = \frac{Q(u)}{W} \frac{1}{\Gamma_1^2 \left\| \gamma_j^2 s_{aj} \right\|_p^{j=1,r}} \tag{3.40}$$

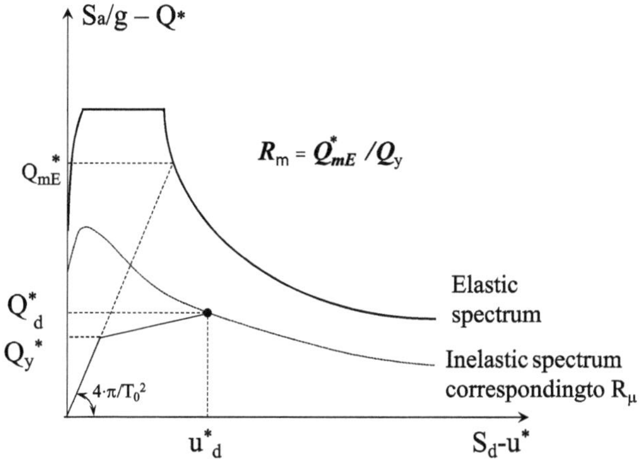

Fig. 3.12 Spectral capacity curve and spectrum demand

(b) Displacements are calculated as

$$u* = \frac{u_i}{\Phi_{i1}} \frac{1}{\Gamma_1 \left\| f_{ij} \gamma_j s_{dj} \right\|_p^{j=1,r}} \tag{3.41}$$

In the formulae presented above, $Q^*(u)$ represents normalized forces, which define the capacity curve, $Q(u)$ are forces determined from the pushover analysis, W is the weight of the structure, u^* are displacements, which define capacity curve, u_i are displacements determined with the pushover analysis, $s_{dj} = S_{dj}/S_{d1}$ is the ratio of the spectral displacement of the mode j and the first mode. Other quantities are described after Eq. 3.39.

The above defined curve can be used together with the inelastic response spectra, presented in the S_a-S_d (acceleration – displacement) format, to obtain the seismic demand of the bridge, following the steps listed below (for more details see Reinhorn (1997) and De Rue (1998) and Fig. 3.12):

(a) First the "elastic force response" $Q^*_{mE} = S_{aE}\, W/g$ is determined from the elastic response spectra for the initial period T_o (defined based on the initial slope of the bi-linear pushover curve)
(b) A "reduction factor", R_μ is calculated from the ratio Q^*_{mE}/Q^*_y (Q^*_y is the yield force defined based on the bi-linear spectral capacity curve)
(c) Then the inelastic spectrum (in S_a–S_d format) is derived by interpolation considering the R_μ determined above
(d) The inelastic response (Q^*_d, and u^*_d) is found at the intersection of the capacity diagram Q^*(u), and the spectrum curve for $R = R_\mu$
(e) For MDOF systems the actual inelastic response demand is then calculated by an inverse application of Eqs. 3.40 and 3.41.

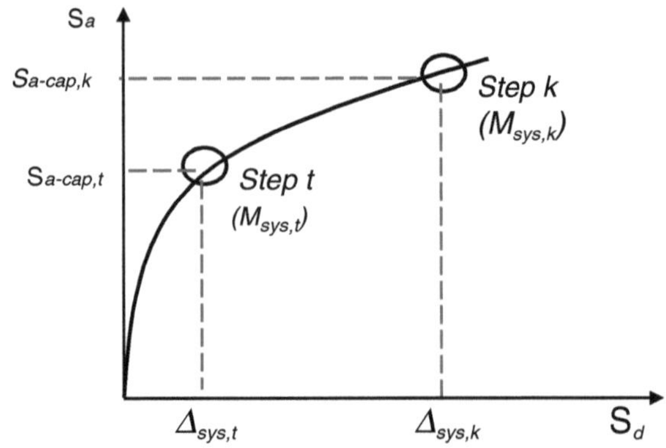

Fig. 3.13 Equivalent SDOF Adaptive Capacity Curve

3.5.3.2 The Adaptive Capacity Spectrum Method (ACSM)

This method combines elements from the Direct Displacement-Based design method (e.g. Priestley and Calvi 2003) and the *Capacity Spectrum Method* (Freeman 1998; ATC 1996), elaborated and revised within an 'adaptive' perspective, for which reason it can also be viewed as an *Adaptive Capacity Spectrum Method (ACSM)*. The procedure is defined as a response spectrum-based approach which employs the substitute structure methodology to model an inelastic system with equivalent elastic properties. The seismic demand is defined by appropriately over-damped elastic response spectra of a given earthquake.

The proposed assessment method for the verification of MDOF bridge structures can be reduced to the following basic steps, explained in detail in what follows: (i) Determination of the 'Equivalent SDOF Adaptive Capacity Curve', (ii) Application of the demand spectrum to the 'Equivalent SDOF Adaptive Capacity Curve', (iii) Determination of the inelastic displacement pattern and of the base shear distribution, (iv) Check of acceptability criteria (pier required strength and displacement).

Description of the Assessment Algorithm

Step 1: Determination of the 'Equivalent SDOF Adaptive Capacity Curve'

The first step is to perform a reliable pushover analysis on a nonlinear model of the MDOF structure. The 'Equivalent SDOF Adaptive Capacity Curve' (Fig. 3.13) is then step-by-step derived by calculating the equivalent system displacement $\Delta_{sys,k}$ and acceleration $S_{a\text{-}cap,k}$ based on the actual deformed shape at each analysis step k, according to Eqs. 3.42 and 3.43, where $V_{b,k}$ is the total base shear of the system and $M_{sys,k}$ is the effective system mass, as defined in Eq. 3.44.

$$\Delta_{sys,k} = \frac{\sum_i m_i \Delta_{i,k}^2}{\sum_i m_i \Delta_{i,k}} \tag{3.42}$$

Fig. 3.14 Individuation of the performance point

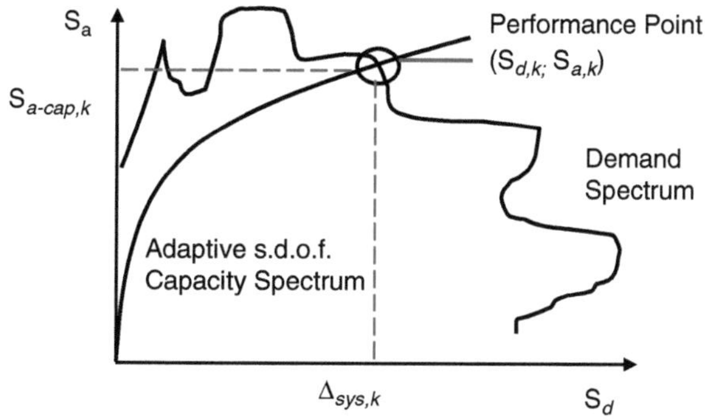

$$S_{a-cap,k} = \frac{V_{b,k}}{M_{sys,k} g} \tag{3.43}$$

$$M_{sys,k} = \frac{\sum_i m_i \Delta_{i,k}}{\Delta_{sys,k}} \tag{3.44}$$

It is noted that $\Delta_{sys,k}$ and $M_{sys,k}$ are defined as the inverse of the Modal Participation Factor and Modal Mass for a modal displacement shape, with the important difference that they are calculated step by step based on the current deformed pattern, rather than on invariant elastic or inelastic modal shape (thus implying that also the $M_{sys,k}$ varies at each step, for which reason the curve is termed 'adaptive').

Step 2: Application of the demand spectrum to the 'Equivalent SDOF Adaptive Capacity Curve'

The developed Adaptive Capacity Curve is intersected with the demand spectrum, providing an estimate of the inelastic acceleration and displacement demand (i.e. performance point) on the structure, as shown in Fig. 3.14. A swift iterative procedure is required at this stage in order to use the appropriate value of equivalent viscous damping to be applied to the demand spectrum: with the performance point obtained with the 10% damped spectrum (any initial value will work, in order to have a starting point), the actual system damping is calculated with the SDOF damping model based, for instance, on the Takeda degrading-stiffness-hysteretic response (Takeda et al. 1970):

$$\zeta_{sys,eff} = 0.05 + \frac{1 - (1 - r)/\sqrt{\mu_{sys}} - r\sqrt{\mu_{sys}}}{\pi} \tag{3.45}$$

In Eq. 3.45, r and μ_{sys} are the post-yielding ratio and the ductility of the SDOF system at the performance point, as calculated by bi-linearising the capacity curve at the performance point, according to the equivalence of areas (i.e. work) between the actual and the bi-linear curve. The procedure is repeated with the spectrum damped with the updated amount of damping and iterated up to the convergence of the damping value: the procedure can be easily implemented in a simple worksheet, and usually converges within two or three iterations.

It is noted that if the demand is described by a real earthquake spectrum rather than a smoothed design spectrum, more than one intersection with the capacity curve may be found. Casarotti and Pinho (2007) have found out that often just one of those intersections provides the convergence with the damping value, and it is generally the intersection corresponding to the largest displacement value. In case more than one intersection converges with the damping, it was found again that the one corresponding to the largest displacement gives results closer to the inelastic response history analysis results. Taking the largest displacement as the performance point would be in any case a reasonably conservative choice, because generally the different intersections have comparable base shear, due to their occurrence in the post-elastic range, but different displacement demand.

Step 3: Determination of the inelastic displacement pattern and of the base shear distribution

Once a performance point on the SDOF capacity curve is established, it suffices to go back to the corresponding level of the pushover database and pick up the actual displacement pattern, base moment and shear values.

Step 4: Check of acceptability criteria (column required strength and displacement)

Having obtained, for each element the force and displacement demand, members are checked to accomplish the shear and the displacement demand.

Differences Between the Capacity Spectrum Method (CSM) and the Adaptive Capacity Spectrum Method (ACSM)

As a recognition of the similarities between the original CSM and the ACSM, an explicit description of their differences is provided herein. The CSM is an iterative procedure essentially applied to buildings, which uses: (i) code-mandated elastic damped acceleration and displacement spectra $S_{\text{a-damp}}/S_{\text{d-damp}}$ or inelastic spectra $S_{\text{a-inel}}/S_{\text{d-inel}}$ (Freeman 1998; Fajfar 1999), (ii) force-based conventional pushover curve, either first- or multi-modal, (iii) equivalent SDOF curve explicitly related to the elastic first mode or to an assumed deformed shape, and based on a modification of the capacity curve built on the displacement of a reference node $\Delta_{\text{reference node}}$, according to the first column of Eq. 3.46, where Γ_1 and M_1^* are respectively the Modal Participation Factor and the Modal Mass of the first mode, $V_{\text{b-pushover}}$ is the total base shear obtained by the pushover analysis, and T the structural period. The ACSM, which has been developed with bridge application in mind (though there is no reason for it not to be applied to buildings as well), makes use of: (i) elastic over-damped spectra, either code-defined or site-specific, (ii) more reliable displacement-based adaptive pushover curves, (iii) equivalent SDOF curve without reference either to any given elastic or inelastic mode shapes, but calculated step by step based on the actual deformed pattern, and not built on a modification of the capacity curve referred to the displacement of a specific physical location. As a consequence, all the 'Equivalent SDOF quantities' (i.e. system displacement $1/\Gamma_{\text{sys}}$ and mass M^*_{sys}), even though of same 'format' of the corresponding modal quantities are also calculated step-by-step based on the actual deformed pattern, according to the second column of Eq. 3.46.

$$\begin{array}{ll} \text{CSM} & \text{ACSM} \\ S_{a/d-demand} = S_{a/d-inel/damp} & S_{a/d-demand} = S_{a/d-damp} \\ S_{a-capacity} = \dfrac{V_{b-pushover}}{M_1^* g} & S_{a-capacity} = \dfrac{V_{b-pushover}}{M_{sys}^* g} \\ S_{d-capacity} = \dfrac{\Delta_{reference\ node}}{\Gamma_1 \varphi_{1,reference\ node}} & S_{d-capacity} = \dfrac{1}{\Gamma_{sys}} \end{array} \tag{3.46}$$

In other words, the ACSM features two types of "adaptiveness". To start with, the pushover analysis algorithm (Antoniou and Pinho 2004b) is fully adaptive, due to the impossibility of a fixed force pattern, characteristic of conventional pushover force-based methods, to accomplish the collapse mode characteristic of a bridge. Indeed, Pinho et al. (2007) showed that the employment of the DAP algorithm leads to better estimates of the inelastic deformed pattern, as well as of the distribution of base forces at a given inelasticity level, independently of structural regularity. The second element of adaptiveness of the method resides in the way the capacity diagram is computed: as stated above, the ACSM can be viewed as an adaptive variant of the CSM approach, because all the 'Equivalent SDOF quantities' vary at each step depending on the current deformed shape, which is not the case in traditional CSM.

References

Abbasi V, Daudeville L, Kotronis P, Mazars J (2004) Using damage mechanics to model a four story RC framed structure submitted to earthquake loading. In: Li et al (eds) Proc 5th Int Conf Fracture Mech Concrete Struct 2:823–830

Alemdar BN, White DW (2005) Displacement, flexibility, and mixed beam-column finite element formulations for distributed plasticity analysis. J Struct Eng 131(12):1811–1819

Antoniou S, Pinho R (2004a) Advantages and limitations of adaptive and non-adaptive force-based pushover procedures. J Earthq Eng 8:497–552

Antoniou S, Pinho R (2004b) Development and verification of a displacement-based adaptive pushover procedure. J Earthq Eng 8:643–661

Antoniou S, Rovithakis A, Pinho R (2002) Development and verification of a fully adaptive pushover procedure. Proceedings of the 12th European Conference on Earthquake Engineering, 9–13 September, London, UK, Paper No. 822

ASCE [American Society of Civil Engineers] (2000) Prestandard and commentary for the seismic rehabilitation of buildings, FEMA 356, Washington D.C., USA

ASCE [American Society of Civil Engineers] (2007) Seismic rehabilitation of existing buildings, ASCE/SEI 41, Washington D.C., USA

ATC [Applied Technology Council] (1996) Seismic evaluation and retrofit of concrete buildings, ATC-40 report, vols 1 and 2, ATC, Redwood City, California, USA

Aydınoğlu MN (2003) An incremental response spectrum analysis based on inelastic spectral displacements for multi-mode seismic performance evaluation. Bull Earthq Eng 1:3–36
Aydınoğlu MN (2004) An improved pushover procedure for engineering practice: incremental response spectrum analysis (IRSA). In: International workshop on performance-based seismic design: concepts and implementation, Bled, Slovenia, PEER Report 2004/05, pp 345–356
Aydınoğlu MN (2005) A code approach for deformation-based seismic performance assessment of reinforced concrete buildings. In: International workshop on seismic performance assessment and rehabilitation of existing buildings, Joint Research Centre (JRC), ELSA Laboratory, Ispra, Italy
Aydınoğlu MN, Fahjan Y (2003) A unified formulation of the piecewise exact method for inelastic seismic demand analysis including the P-delta effect. Earthq Eng Struct Dyn 32:871–890
Aydınoğlu MN, Önem G (2009) Nonlinear performance assessment of bridges with incremental response spectrum analysis (IRSA) procedure. In: Papadrakakis M et al (eds) Computational structural dynamics and earthquake engineering. CRC Press, Leiden, pp 393–400
Aydınoğlu MN, Önem G (2010) Evaluation of analysis procedures for seismic assessment and retrofit design, Chapter 8. In: Earthquake engineering in Europe. Springer, Netherlands
Biggs JM (1964) Structural dynamics. McGraw-Hill, New York
Bracci JM, Kunnath SK, Reinhorn AM (1997) Seismic performance and retrofit evaluation for reinforced concrete structures. J Struct Eng 123(1):3–10
Casarotti C, Pinho R (2007) An adaptive capacity spectrum method for assessment of bridges subjected to earthquake action. Bull Earthq Eng 5(3):377–390
CEN [Comité Européen de Normalisation] (2004) Eurocode 8: Design of structures for earthquake resistance – Part 1: General rules, seismic actions and rules for buildings, Brussels
CEN [Comité Européen de Normalisation] (2005) Eurocode 8: Design provisions of structures for earthquake resistance – Part 2: Bridges, Brussels
Chopra AK (2001) Dynamics of structures: theory and applications to earthquake engineering, 2nd edn. Prentice Hall, Englewood Cliffs
Chopra AK, Goel RK (1999) Capacity-Demand-Diagram methods based on inelastic design spectrum. Earthq Spectra 15(4):637–656
Chopra AK, Goel RK (2002) A modal pushover analysis procedure for estimating seismic demands for buildings. Earthq Eng Struct Dyn 31(3):561–582
Clough RW, Penzien J (1993) Dynamics of structures, 2nd edn. McGraw Hill, New York
De Rue GM (1998) Nonlinear static procedure analysis of 3D structures for design applications. Master thesis, University of New York at Buffalo
Elnashai AS (2001) Advanced inelastic static (pushover) analysis for earthquake applications. Struct Eng Mech 12(1):51–69
Fajfar P (1999) Capacity spectrum method based on inelastic demand spectra. Earthq Eng Struct Dyn 28(9):979–993
Fajfar P (2000) A nonlinear analysis method for performance-based seismic design. Earthq Spectra 16(3):573–592
Fajfar P (2007) Seismic assessment of structures by a practice oriented method. In: Ibrahimbegovic A, Kozar I (eds) Extreme man-made and natural hazards in dynamics of structures, NATO security through science series, Springer Netherlands
Fajfar P, Fischinger M (1987) Non-linear seismic analysis of RC buildings: implications of a case study. Eur Earthq Eng 1:31–43
Fajfar P, Fischinger M (1988) N2- a method for non-linear seismic analysis of regular buildings. Proc 9th World Conf Earthq Eng Tokyo V:111–116
FEMA [Federal Emergency Management Agency] (1997) NEHRP guidelines for the seismic rehabilitation of buildings, FEMA-273, Washington D.C.
FEMA [Federal Emergency Management Agency] (2005) Improvement of nonlinear static seismic analysis procedures (ATC-55/FEMA 440), Washington D.C.
FEMA [Federal Emergency Management Agency] (2009) Effects of strength and stiffness degradation on seismic response, FEMA P440A, Washington D.C.

Filippou FC, Fenves GL (2004) Methods of analysis for earthquake-resistant structures. In: Bozorgnia Y, Bertero VV (eds) Earthquake engineering – from engineering seismology to performance-based engineering. Cambridge University Press, Cambridge
Fischinger M, Beg D, Isaković T, Tomazevic M, Zarnic R (2004) Performance based assessment – from general methodologies to specific implementations. In: Proceedings of the International workshop on performance based seismic design, Bled, Slovenia 2004; published in PEER report 2004-05 (UC Berkeley), pp 293–308
Fragiadakis M, Papadrakakis M (2008) Modeling, analysis and reliability of seismically excited structures: computational issues. Int J Comput Method 5(4):483–511
Freeman SA (1998) Development and use of capacity spectrum method. In: Proceedings of the 6th U.S. national conference on earthquake engineering, Seattle, CD-ROM, EERI, Oakland
Freeman SA, Nicoletti JP, Tyrell JV (1975) Evaluations of existing buildings for seismic risk – a case study of Puget Sound Naval Shipyard, Bremerton, Washington. In: Proceedings of the 1st U.S. national conference on earthquake engineering, Oakland, California, pp 113–122
Freitas JAT, Almeida JPM, Pereira EMBR (1999) Non-conventional formulations for the finite element method. Comput Mech 23:488–501
Goel RK, Chopra AK (2004) Extension of modal pushover analysis to compute member forces. Earthq Spectra 21(1):125–139
Hall JF (2006) Problems encountered from the use (or misuse) of Rayleigh damping. Earthq Eng Struct Dyn 35:525–545
Han SW, Chopra AK (2006) Approximate incremental dynamic analysis using the modal pushover analysis procedure. Earthq Eng Struct Dyn 35(15):1853–1873
Hellesland J, Scordelis A (1981) Analysis of RC bridge columns under imposed deformations. IABSE Colloquium, Delft, pp 545–559
Isaković T, Fischinger M (2006) Higher modes in simplified inelastic seismic analysis of single column bent viaducts. Earthq Eng Struct Dyn 35(1):95–114
Isaković T, Fischinger M (2011) Applicability of pushover methods to the seismic analyses of an RC bridge, experimentally tested on three shake tables. J Earthq Eng 15(2):303–320
Isaković T, Lazaro Nino MP, Fischinger M (2008) Applicability of pushover methods for the seismic analysis of single column bent viaducts. Earthq Eng Struct Dyn 37(8):1185–1202
Kappos AJ, Petranis C (2001) Reliability of pushover analysis – based methods for seismic assessment of RC buildings. In: Corz A, Brebbia CA (eds) Earthquake resistant engineering structures III. WIT Press, Southampton, pp 407–416
Lupoi A, Franchin P, Pinto PE (2007) Further probing of the suitability of push-over analysis for the seismic assessment of bridge structures. In: Proceedings of the ECCOMAS thematic conference on computational methods in structural dynamics and earthquake engineering (COMPDYN), Rethymno, Greece, Paper No. 1045
Mari A, Scordelis A (1984) Nonlinear geometric material and time dependent analysis of three dimensional reinforced and prestressed concrete frames, SESM Report 82-12, Department of Civil Engineering, University of California, Berkeley
Ministry of Public Works and Settlement (2007) Specification for buildings to be built in earthquake zones. Ankara, Turkey (in Turkish)
Ministry of Transportation (2008) Seismic technical specification for coastal and harbour structures, railways and airport construction – Chapter 3. In: Design essentials of railway bridges under earthquake action. Ankara, Turkey (in Turkish)
Miranda E, Akkar SD (2003) Dynamic instability of simple structural systems. J Struct Eng 129:1722–1726
Neuenhofer A, Filippou FC (1997) Evaluation of nonlinear frame finite-element models. J Struct Eng 123(7):958–966
OASP [Earthquake Protection Organization of Greece] (2009) Code for structural interventions (KANEPE), Final draft, Athens. http://www.oasp.gr/ (in Greek)

Paraskeva TS, Kappos AJ (2008) An improved modal pushover analysis procedure for the seismic assessment of bridges. In: Proceedings of the 14th world conference on earthquake engineering (14WCEE), Beijing, China, Paper No. 14-0236

Paraskeva TS, Kappos AJ (2010) Further development of a multimodal pushover analysis procedure for seismic assessment of bridges. Earthq Eng Struct Dyn 39(2):211–222

Paraskeva TS, Kappos AJ, Sextos AG (2006) Extension of modal pushover analysis to seismic assessment of bridges. Earthq Eng Struct Dyn 35(11):1269–1293

Paret TF, Sasaki KK, Eilbeck DH, Freeman SA (1996) Approximate inelastic procedures to identify failure mechanisms from higher mode effects. In: Proceedings of 11th world conference on earthquake engineering, Acapulco, Mexico, Paper No. 966

Pegon P (1996) Derivation of consistent proportional viscous damping matrices. JRC Research Report No. I.96.49, Ispra, Italy

Penelis GrG, Kappos AJ (2005) Inelastic torsion effects in 3D pushover analysis of buildings. In: Proceedings of the 4th European workshop on the seismic behaviour of irregular and complex structures, Thessaloniki, Greece, Paper No. 51

Pinho R, Casarotti C, Antoniou S (2007) A comparison of single-run pushover analysis techniques for seismic assessment of bridges. Earthq Eng Struct Dyn 36(10):1347–1362

Priestley MJN, Calvi GM (2003) Direct displacement-based seismic design of concrete bridges. In: Proceedings of the 5th ACI international conference of seismic bridge design and retrofit for earthquake resistance, December 8–9, 2003, La Jolla, California

Priestley MJN, Grant DN (2005) Viscous damping in seismic design and analysis. J Earthq Eng 1(9):229–255

Reinhorn AM (1997) Inelastic analysis techniques in seismic evaluations. In: Proceedings of the international workshop on seismic design methodologies for the next generation of codes, Bled, Slovenia, pp 277–288

Saiidi M, Sozen M (1981) Simple nonlinear seismic analysis of RC structures. J Struct Div ASCE 107(ST5):937–952

Sasaki KK, Freeman SA, Paret TF (1998) Multimode pushover procedure (MMP): a method to identify the effects of higher modes in a pushover analysis. In: Proceedings of 6th U.S. national conference on earthquake engineering, Seattle, Washington

Spacone E, Ciampi V, Filippou FC (1996) Mixed formulation of nonlinear beam finite element. Comput Struct 58(1):71–83

Takeda T, Sozen M, Nielsen N (1970) Reinforced concrete response to simulated earthquakes. J Struct Div ASCE 96(12):2557–2573

Vamvatsikos D, Cornell CA (2002) Incremental dynamic analysis. Earthq Eng Struct Dyn 31(3):491–514

Wakabayashi M (1986) Design of earthquake-resistant buildings. McGraw-Hill, New York

Wilson E (2001) Static and dynamic analysis of structures, Computers and Structures Inc, Berkeley, California. (excerpts available at URL: www.csiberkeley.com/support_technical_papers.html)

Chapter 4
Case Studies and Comparative Evaluation of Methods

Tatjana Isaković, Antonio Arêde, Donatello Cardone, Pedro Delgado, Matej Fischinger, Andreas J. Kappos, Nelson V. Pouca, Rui Pinho, and Anastasios Sextos

4.1 Introduction

Different simplified nonlinear methods, presented in Chap. 3, are based on different assumptions, which define the level of their sophistication and limit their applicability. The more sophisticated methods have usually broader scope of applicability; however none of them is universal. Therefore, each user has to be aware of the

T. Isaković (✉) • M. Fischinger
Faculty of Civil and Geodetic Engineering, University of Ljubljana, Jamova 2, 1000 Ljubljana, Slovenia
e-mail: tisak@ikpir.fgg.uni-lj.si; matej.fischinger@ikpir.fgg.uni-lj.si

A. Arêde • N.V. Pouca
Departamento de Engenharia Civil, Faculdade de Engenharia, University of Porto, Rua Dr. Roberto Frias, s/n, 4200-465 Porto, Portugal
e-mail: aarede@fe.up.pt; nelsonvp@fe.up.pt

D. Cardone
Department of Structures, Geotechnics and Applied Geology, University of Basilicata, Viale dell'Ateneo Lucano 10, 85100 Potenza, Italy
e-mail: donatello.cardone@unibas.it

P. Delgado
Escola Superior de Tecnologia e Gestão, Instituto Politécnico de Viana do Castelo, Avenida do Atlântico, 4900-348 Viana do Castelo, Portugal
e-mail: pdelgado@estg.ipvc.pt

A.J. Kappos • A. Sextos
Department of Civil Engineering, Aristotle University of Thessaloniki, 54124 Thessaloniki, Greece
e-mail: ajkap@civil.auth.gr; asextos@civil.auth.gr

R. Pinho
Department of Structural Mechanics, University of Pavia, Via Ferrata 1, 27100 Pavia, Italy
e-mail: rui.pinho@unipv.it

A.J. Kappos et al. (eds.), *Seismic Design and Assessment of Bridges: Inelastic Methods of Analysis and Case Studies*, Geotechnical, Geological and Earthquake Engineering 21, DOI 10.1007/978-94-007-3943-7_4, method's limitations before he/she chooses it for the analysis of a certain type of bridge. This chapter is intended to assist the reader in selecting an appropriate method.

In Sect. 4.2 the main parameters that influence the applicability of non-linear pushover based methods are summarized. They are related to the length of the superstructure, its stiffness, the stiffness and the strength of columns, as well as the type of supports above the abutments. All these important issues are discussed and illustrated through typical examples.

The main parameters that influence the applicability of different methods were identified and analyzed within several extensive parametric studies. These studies are presented in Sect. 4.3. Analyzed bridges can be divided into two main groups: (a) those, which meet the majority of the requirements of the modern seismic design codes (Sects. 4.3.1,–4.3.3), and (b) older bridges, which were built before the basic principles of modern seismic design philosophy were established (Sect. 4.3.4). Thus the presented case studies encompass a broad range of different types of bridges. The first group includes primarily single-column bent, straight and curved, viaducts, with different types of bearings at the abutments, and different types of connections between superstructure and columns (monolithic or pinned). The second group of bridges consists of multi-span simply-supported bridges.

The applicability and accuracy of different pushover methods was analyzed mostly based on the comparison with nonlinear response history analysis (NRHA). The differences with respect to NRHA were in many cases defined qualitatively, but some attempts were also made to quantify them. Different quantities (displacements, plastic rotations, shear forces in columns, etc.), reported in Sects. 4.3.1,–4.3.4, were used to measure them.

Two indices that can be used to quantify the applicability of different pushover methods are presented in Sects. 4.3.2 (Index of applicability) and 4.3.3 (Bridge Index – BI).

NRHA is in the majority of cases considered as the most accurate method. Therefore it is often used as a reference method to test the accuracy of other, simplified, non-linear methods. Since it also includes certain assumptions, this method, too, provides only an approximation of the actual response.

To test the accuracy of NRHA and to compare different pushover-based methods, a special case study involving an experimentally tested bridge was performed. The tested structure was a large-scale model of the typical two-column, two-span frame-unit of common concrete bridges in the USA. However it can be also identified as a unit of some types of bridges in other parts of the world. It was tested on three shake tables at the University of Nevada, Reno, USA, under different seismic intensity levels. The results of this test and correlation with different analytical methods are presented in Sect. 4.4.1.

Section 4.4.2 comprises experimental evaluation of NRHA using an example of an existing bridge with hollow box columns with substandard construction details.

4.2 Basic Parameters That Influence the Applicability of Pushover Methods

The range of applicability of each pushover-based method depends on its assumptions. The basic assumptions are mostly related to the influence of higher modes on the response, and their changes under different seismic intensities. For example, different single-mode pushover methods, which are adopted by many modern codes (see Chap. 3) are based on the assumption that the response is governed by only one mode of vibration, which has a shape that does not depend on the seismic intensity. This is usually the case in the longitudinal direction of the majority of bridges with continuous deck.

Quite often, the response of bridges in the transverse direction is more complex than in the longitudinal direction. Hence, there are many structures where the influence of higher modes in the transverse direction is considerable, and moreover where these modes considerably change, under different seismic intensities. Usually, the standard single mode pushover methods are not accurate enough in such cases. Hence either the multimode pushover methods or NRHA should be employed for their analysis.

Since very often the more accurate pushover analysis is not needed in the longitudinal direction, the discussions presented in Sects. 4.3.1,–4.3.3 and 4.4.1, as well as that presented below are mostly devoted to the transverse direction of bridges. In the following paragraphs the key parameters that influence the response of bridges are addressed. These parameters were used to categorize bridges into several groups depending on the complexity of their response. For each of these groups, guidelines for the choice of an appropriate type of analysis are provided in Chap. 5 of this book and are summarized in Table 5.1.

Although the discussion in this chapter mostly focuses on the transverse direction, the response and vulnerability in the longitudinal direction is also analyzed, in the example of existing bridges, which usually have quite complex response in their longitudinal direction as well (Sect. 4.3.4).

The influence of higher modes and their changes under different seismic intensities depend mainly on: (a) the ratio of the superstructure stiffness to the stiffness of the bents (the length of the bridge and number and location of short columns along the bridge), (b) the relative strength of the columns, related to the seismic intensity, and (c) boundary conditions at the abutments (mostly in short bridges).

In bridges where the superstructure is considerably stiffer than supporting elements (piers) the influence of the higher modes is in general negligible. Typical representatives are short and medium span bridges (i.e. the length of a bridge is less than about 500 m), which are not supported by very stiff (very short) piers (e.g. in single-column piers the height of columns exceeds 10 m). An example is presented in Fig. 4.1. In general the standard pushover methods can be used in such bridges (see also Chap. 5), unless the predominant mode considerably changes when the seismic intensity is changed.

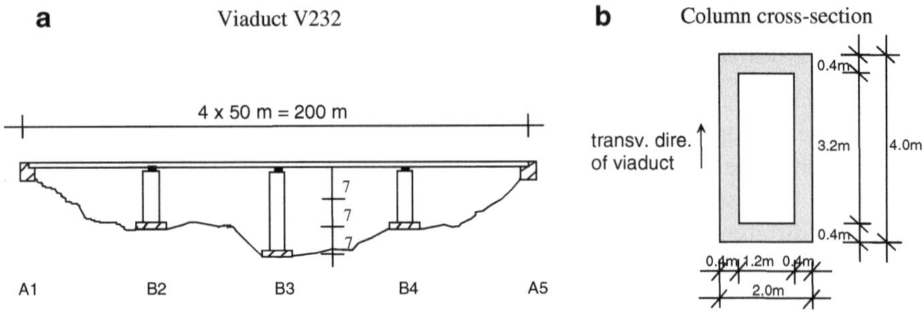

Fig. 4.1 An example of a bridge where the response is influenced by only one predominant mode, which does not change considerably when the seismic intensity is changed. (**a**) Viaduct V232, (**b**) Column cross-section

Considerable changes of the predominant mode can be expected (and were observed) first of all in short bridges, where the displacements of the superstructure above the abutments are not restrained. In such bridges, the predominant mode usually changes considerably, when the damage of the side columns reduces their stiffness to such extent, that the torsional stiffness of the bridge becomes lower than its translational stiffness (the definition of torsional and translational stiffness of the bridge can be found in Isaković et al. 2003). An example is presented in Fig. 4.2, while another one is given later (Fig. 4.6); the response of the bridge presented in Fig. 4.2 is analyzed and discussed in more detail in Sect. 4.4.1.

The same structure (Fig. 4.2) can be also used to illustrate the importance of the relative strength of columns to the response. Before substantial yielding of the side columns occurs, their stiffness is large enough to provide torsional stiffness of the structure, which is larger than the translational one. This is ensured at lower seismic intensities. Consequently, the single mode methods give quite accurate results.

However, when the seismic intensity is considerably increased, the effective stiffness of side columns can be considerably reduced. Consequently, the torsional stiffness of the bridge becomes gradually smaller than the translational stiffness. The predominant mode (as explained above), therefore, can be considerably changed and the results of single mode methods can be considerably unconservative compared to the NRHA and the actually recorded response.

In some bridges (similar to the bridge presented in Fig. 4.3), the opposite trend has been observed. In bridges with short central columns the important influence of higher modes can be observed, particularly for the lower seismic intensities. The main reason is the large relative stiffness of columns, compared to the stiffness of the superstructure. In this type of bridges, due to the initially large stiffness of columns, the superstructure is not stiff enough to dominate the response of the entire bridge. Each column tends to move in its own way. Therefore, in such bridges, the higher modes are important at lower seismic intensities.

However, when the seismic intensity is increased to a certain level, the columns start to yield and their stiffness is reduced. Particularly important is the substantial

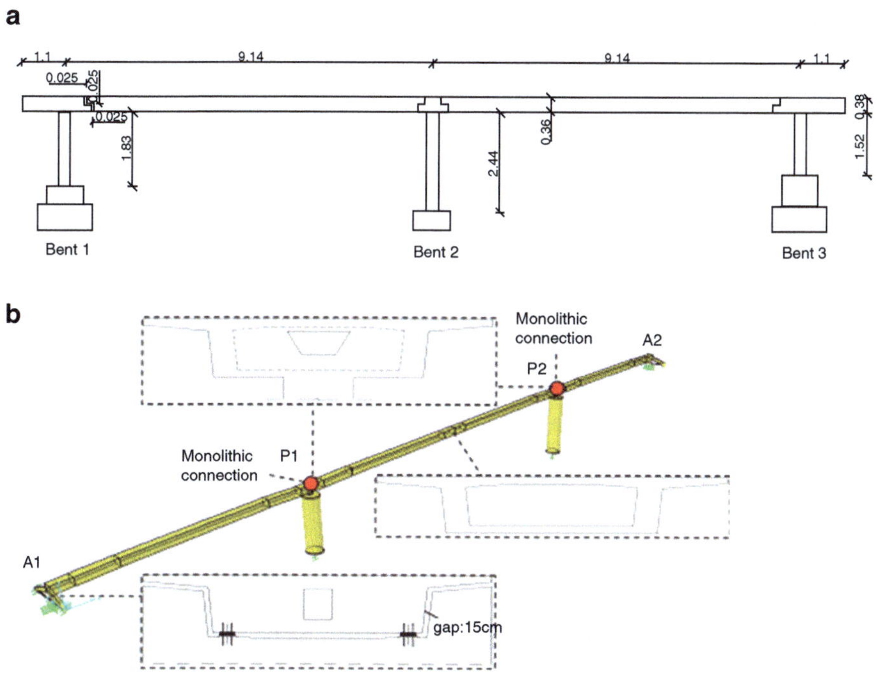

Fig. 4.2 An example of a bridge where the response is influenced by one predominant mode, which considerably changes when the seismic intensity is increased

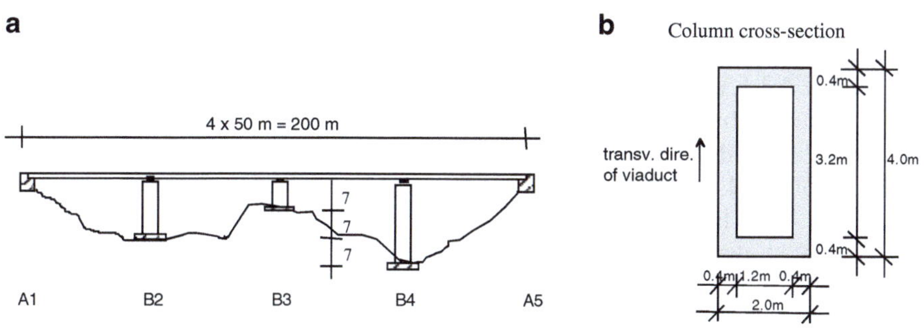

Fig. 4.3 An example of a bridge where the response is significantly influenced by higher modes unless yielding in the central column is reached. (**a**) Viaduct V213, (**b**) Column cross-section

reduction of the effective stiffness of central short columns. Due to this reduction, the substructure assumes the more important role and it is able to enforce the way of movement to columns. Consequently, only one mode becomes important and the single mode methods become more accurate at higher seismic intensities.

In long bridges (see Fig. 4.4) the influence of higher modes in the majority of cases does not depend on the stiffness of columns and their strength. In common

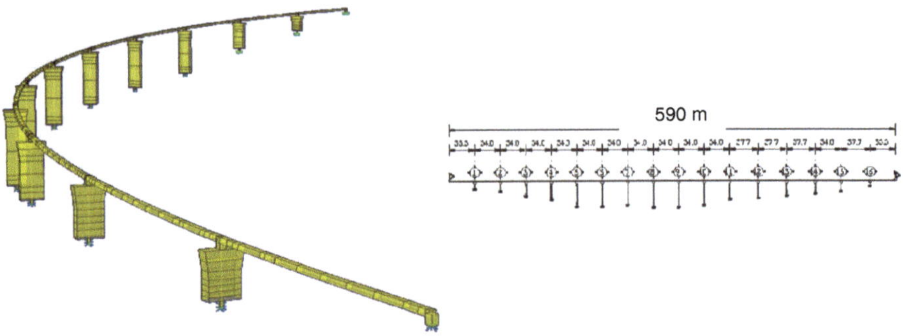

Fig. 4.4 The response of long and/or curved bridges is typically influenced by the higher modes regardless of the stiffness of columns and the seismic intensity

bridges of this type the superstructure becomes relatively flexible, and consequently the higher modes become important regardless of the stiffness and strength of columns.

4.3 Case Studies – Comparison of Alternative Methods

In this section parametric studies of different types of bridges are presented. Structures were mostly analyzed in the transverse direction. The majority of the examples were single column bent viaducts. They differed regarding the influence of higher modes on the response, as well as their variability. Sections 4.3.1, 4.3.2 and 4.3.3 include case studies of bridges that were designed according to modern codes.

In Sect. 4.3.1, three as-built single column bent bridges were analyzed: (a) a long 12-span curved bridge supported by columns having heights in the range between 11 and 27 m, and (b) two three-span bridges of 100 m (typical overpass with two columns of 8 and 10 m height) and of 247 m total length (supported on two columns of 36 and 45 m height). In this study the standard single mode pushover method (named SPA), the improved MPA method, and NRHA were compared.

In Sects. 4.3.2 and 4.3.3 similar types of four-span relatively short bridges of 200 m total length were analyzed. They were supported on columns with different configuration, having height 7, 14 and 21 m. Some of the examples in these two studies seem to be the same, since they have the same geometry. However, the seismic response of these bridges was in some cases different, since they were designed taking into account different seismic intensities and different design code requirements. Consequently, column reinforcement and their nonlinear response were sometimes different (see Chap. 2 for more details).

The examples differ also regarding the modelling of abutments. In Sect. 4.3.3 bridges with two different types of abutments were analyzed: (i) continuous deck-abutment connections supported on piles, with a bilinear behaviour, and (ii) deck ends supported on pot bearings. Bridges, analyzed in Sect. 4.3.2, were mostly pinned at the abutments, except the highly irregular bridges, where rollers were considered above the abutments. In both studies the analysis of short bridges was extended to long structures, where the length of the superstructure exceeds 500 m.

In Sect. 4.3.2 three pushover methods, the N2 method, the MPA, and the IRSA were compared with NRHA and experimental results. In Sect. 4.3.3 the applicability of four methods, CSM, N2, MPA, and ACSM was evaluated also through NRHA.

Two indices that can be used to quantify the applicability of different pushover methods for the analysis of bridges are presented in Sects. 4.3.2 (Index of applicability) and 4.3.3 (Bridge Index – BI).

In Sect. 4.3.4 older bridges that do not meet the criteria of modern codes were analyzed using the IACSM method, which was evaluated against NRHA. The analysis of nine structures that differ with respect to the total length of the bridge, and the heights and configurations of columns is presented. All of the analyzed structures were multi-span simply-supported bridges. The performed analysis is illustrated in the example of a typical five-span structure with span length of 33.5 m and column heights in the range from 4.5 to 10 m. A risk assessment of the analyzed bridges is also presented.

4.3.1 *Case Study 1: Single-mode, Multimodal Pushover, and Dynamic Response History, Analyses of Bridges*

4.3.1.1 Description of Studied Bridges

In order to investigate the accuracy and also the practicality of the improved multimodal pushover analysis procedure described in Sect. 3.5.1, it was deemed appropriate to apply it to three actual concrete bridges; all of them are part of the Egnatia Motorway in N. Greece, and they are described in more detail in Moschonas et al. (2009).

The Krystallopigi bridge (Fig. 4.5) is a 12-span structure of 638 m total length. The curvature in plan (radius equal to 488 m) of the bridge adds to the expected complexity of its dynamic behaviour. Piers are rectangular hollow reinforced concrete members, while the height of the 11 piers varies between 11 and 27 m. For the end piers P1–P3 and P9–P11 (Fig. 4.5) a bearing type pier-to-deck connection is adopted, while the interior (taller) piers are monolithically connected to the deck.

The second structure is an overpass (overcrossing) bridge with three spans (denoted hereafter as “overpass bridge”) and total length equal to 100 m, typical in modern motorway construction (Fig. 4.6). Piers have a cylindrical cross section, a common choice for bridges both in Europe and in other areas, while the pier heights are 8 and 10 m. The deck is monolithically connected to the piers, while it rests on its two abutments through elastomeric bearings; movement in both the

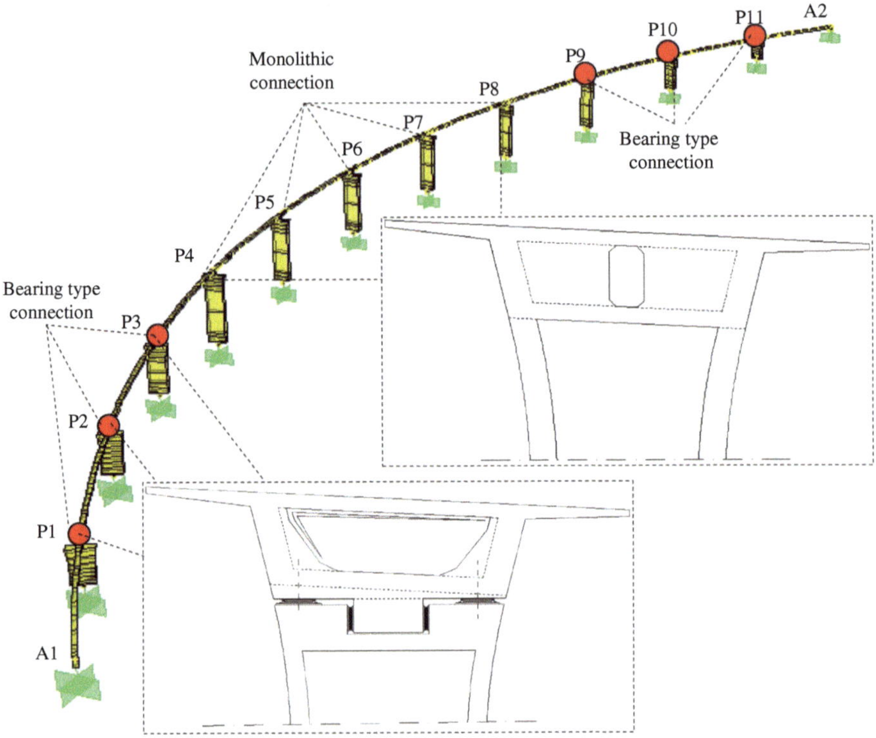

Fig. 4.5 Layout of Krystallopigi bridge finite element modelling

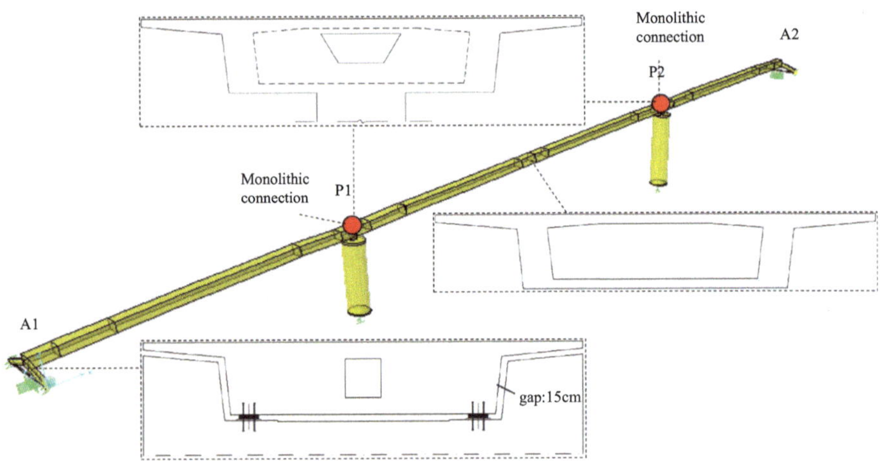

Fig. 4.6 Layout of the overpass bridge finite element modelling

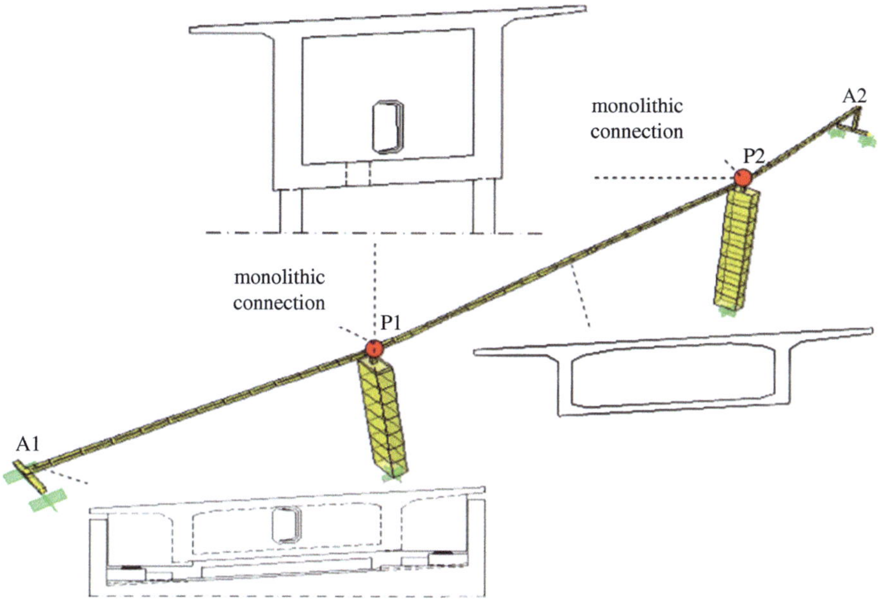

Fig. 4.7 Finite element modelling of the G11 bridge

longitudinal and the transverse directions is initially allowed at the abutments, but transverse displacements are restrained whenever the 15 cm gap shown at the bottom of Fig. 4.6 is closed.

The third structure is a three-span ravine bridge of 247 m total length, with small curvature in plan (Fig. 4.7) (denoted as the "G11 bridge"). Piers are rectangular hollow RC members, while the pier heights are 36 and 45 m. Both piers are monolithically connected to the deck (cantilever construction).

The Greek Seismic Code EAK2000 (Ministry of Public Works 2003) design spectrum scaled to 0.24 g for the first bridge and to 0.16 g for the second and third ones (different seismic zones), was used for seismic design. The design spectrum corresponded to ground category 'B' of EAK (same as in the ENV version of Eurocode 8, closer to 'C' in the final version of the Code (CEN 2004)). All bridges were designed as ductile structures (plastic hinges expected in the piers); behaviour factors q = 3.0 for the Krystallopigi bridge and the G11 bridge and q = 2.4 for the overpass, were adopted for design. The bridges were assessed using standard pushover analysis (first mode loading), pushover analysis for a 'uniform' loading pattern (as required by Eurocode 8 (CEN 2004) and by the ASCE Standard 41-07 (ASCE/SEI 2007)), modal pushover analysis as proposed in Paraskeva et al. (2006), and improved modal pushover analysis as described in Sect. 3.5.1; the demand spectrum in all analyses was the design one or multiples of it. The bridges were subsequently assessed using nonlinear response-history analysis (NRHA), for artificial records closely matching the demand spectrum, derived using the ASING program (Sextos et al. 2003).

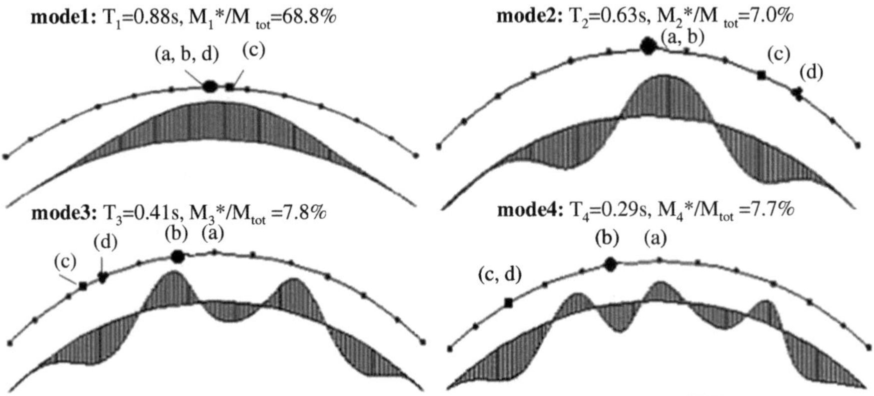

	mode 1	mode 2	mode 3	mode 4
$x_{centre\ mass}$/(total length)	0.50	0.50	0.50	0.50
x_{SDOF}/(total length)	0.50	0.52	0.44	0.42
x_{max}/(total length)	0.54	0.80	0.19	0.14
$x_{critical\ pier}$/(total length)	0.50	0.86	0.23	0.14

Fig. 4.8 Modal force distribution, location of the equivalent *SDOF* systems, and modal parameters for the main transverse modes of Krystallopigi bridge

All inelastic analyses were carried out using the SAP2000 software package (CSI 2003). Plastic hinging in the piers had to be modelled slightly differently in the NRHA and the pushover analysis, due to limitations of the software used. More specifically, non-linear rotational spring elements were used in the finite element models used in NRHA, while the built-in beam hinge feature of SAP2000 was implemented in the models set up for pushover analysis. In both cases, though, the same moment-rotation (M-θ) relationship was used (i.e. bilinear with 2–6% hardening, depending on the calculated ultimate moment), with input parameters defined from fibre analysis performed for each pier section, utilising the computer program RCCOLA-90 (Kappos 2002) that includes the confinement model for concrete described in Kappos (1991).

4.3.1.2 Non Linear Static Analyses

Fundamental mode-based ('standard') pushover analyses as well as a 'uniform' loading pushover analysis were first performed for assessing the inelastic response of the selected bridges; results of these analyses (reported only briefly herein, due to space limitations) were presented in detail elsewhere (Paraskeva et al. 2006; Kappos and Paraskeva 2008).

The dynamic characteristics of the bridges, required within the context of the MPA approach, were determined using standard eigenvalue analysis. Figure 4.8 illustrates the first four transverse mode shapes of Krystallopigi bridge, while

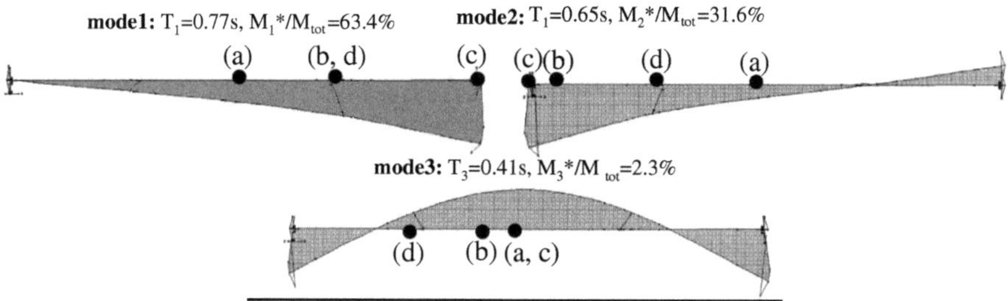

		mode1	mode2	mode3
$x_{centre\ mass}$/(total length)	(a)	0.50	0.50	0.50
x_{SDOF}/(total length)	(b)	0.73	0.08	0.44
x_{max}(total length)	(c)	1.00	0.00	0.50
$x_{critical\ pier}$/(total length)	(d)	0.73	0.27	0.27

Fig. 4.9 Modal force distribution, location of the equivalent *SDOF* systems, and modal parameters for the main transverse modes of the overpass bridge

Fig. 4.9 illustrates the first three transverse mode shapes of the overpass bridge, together with the corresponding participation factors and mass ratios, as well as the locations of monitoring points for each mode; the G11 bridge (Fig. 4.7) shows similar mode shapes as the Krystallopigi bridge. Consideration of the modes shown in Figs. 4.8 and 4.9 assures that more than 90% of the total mass in the transverse direction is considered. Applying the modal load pattern of the nth mode in the transverse direction of the bridge, the corresponding pushover curve was constructed and then idealized as a bilinear curve.

As noted under Step 4 of the MPA procedure (see Sect. 3.5.1), the inelastic spectra-based version of CSM was used for defining the displacement demand for a given earthquake intensity. Figures 4.10 and 4.11 illustrate the deck displacements of the selected bridges derived using pushover analysis for each mode independently, as well as the MPA procedure initially proposed in Paraskeva et al. (2006). If the structure remains elastic for the given earthquake intensity, both spectral displacement S_d and the product $\Gamma_n \cdot \phi_n$ will be independent of the selection of the control (monitoring) point; this means that deck displacements are independent of the location of the monitoring point. On the contrary, it was found that deck displacements derived with respect to different control points, for inelastic behaviour of the structure are not identical but rather the estimated deformed shape of the bridge depends on the monitoring point selected for drawing the pushover curve for each mode.

For inelastic behaviour, it appears that Eq. 3.2 in Sect. 3.5.1 gives a different value of u_{cn}, not only because of the deviation of the elastic mode shape ϕ_n from the actual deformed shape of the structure, but also due to the fact that the spectral displacement S_d is dependent on the selection of monitoring point if the structure exhibits inelastic behaviour (due to the bilinearization of the capacity curve). An improved target displacement of the monitoring point is calculated (from Eq. 3.3) using ϕ_n', the actual deformed shape of the structure

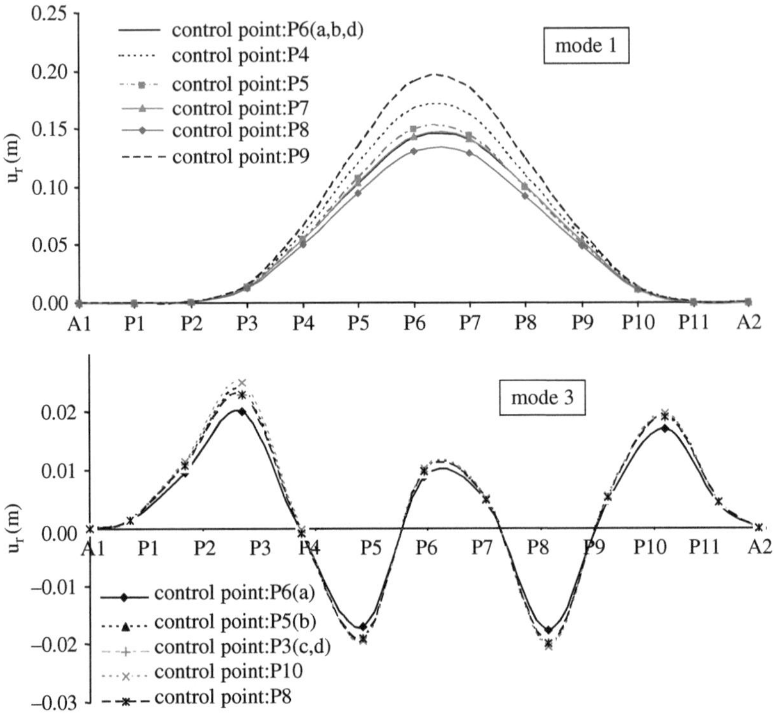

Fig. 4.10 'Modal' deck displacements derived with respect to different control points – inelastic behaviour of Krystallopigi bridge ($A_g = 0.32$ g)

(see Figs. 4.10 and 4.11), while the spectral displacement remains the same. The response quantities of interest are evaluated by extracting from the 'database' the values of the desired responses, r_n, for the analysis step at which the displacement at the control point is equal to $u_{cn'}$ (the improved estimate of u_{cn} derived on the basis of $\boldsymbol{\phi}_n'$). Figures 4.12 and 4.13 illustrate the deck displacements of Krystallopigi bridge and overpass bridge, respectively, calculated from SPA using $u_{cn'}$ as target displacement for each mode. It is noted that, due to the approximations involved in the CSM procedure, deck displacements derived with respect to different control points are not the same, but differences are significantly reduced and results are deemed acceptable for all practical purposes.

From Figs. 4.10,–4.13 it is observed that the differences between deck displacements derived with respect to different control points, as well as the improvement in the prediction of deck displacements using the procedure proposed here, are more significant in the case of Krystallopigi bridge than in the overpass bridge. This is attributed to the larger length combined with the curvature in plan of the former bridge, which amplifies the complexity of its dynamic behaviour and renders more significant the contribution of higher modes.

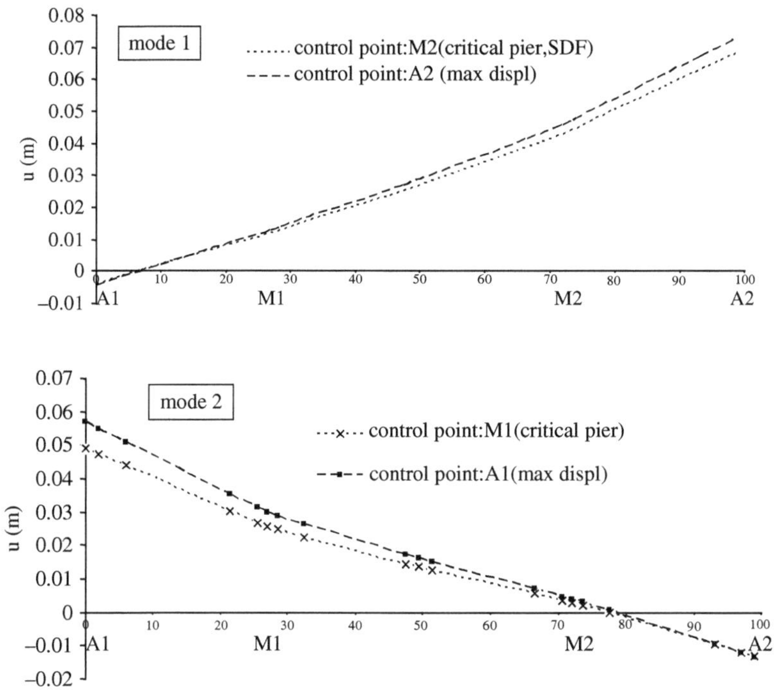

Fig. 4.11 'Modal' deck displacements derived with respect to different control points – *inelastic* behaviour of the overpass bridge ($A_g = 0.16$ g)

4.3.1.3 Evaluation of Different Procedures

Results of the standard and modal pushover approaches were evaluated by comparing them with those from non-linear response history analysis, the latter considered as the most rigorous procedure to estimate seismic demand. To this effect, a series of NRHAs was performed using five artificial records compatible with the EAK2000 elastic spectrum and generated using the computer code ASING (Sextos et al. 2003). The Newmark $\gamma = 1/2$, $\beta = 1/4$ integration method was used, with time step $\Delta t = 0.0025$ s and a total of 10,000 steps (25 s of input). A uniform damping value of 5% was assumed for all modes of vibration, while hysteretic damping was accounted for through the elastoplastic behaviour of the structural members.

The displacements determined by the SPA and MPA procedures were compared to those from NRHA for increasing levels of earthquake excitation, as shown in Figs. 4.11, 4.12 and 4.14. It is noted that the deck displacements shown in the figures as the NRHA case are the average of the peak displacements recorded in the structure during the five response-history analyses. Besides, in this and all subsequent figures, the displacement demand is estimated independently in static

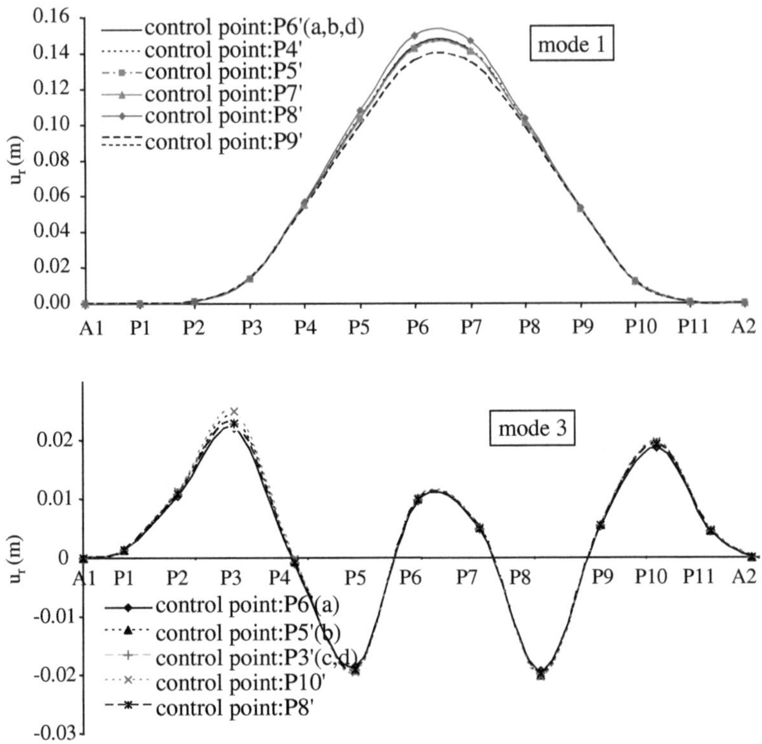

Fig. 4.12 'Modal' deck displacements derived with respect to different control points using $u_{rn'}$ as target displacement according to the *improved MPA* procedure- Krystallopigi bridge ($A_g = 0.32$ g)

and dynamic (response-history) inelastic analysis, whereas in some previous studies comparisons of displacement profiles are made assuming the same maximum displacement in both cases; the choice adopted here is deemed as more relevant for practical applications, as it permits an evaluation of all aspects of the proposed procedure.

In the case of Krystallopigi bridge (Fig. 4.14) it is observed that the SPA procedure predicts well the maximum transverse displacements only in the area of the central piers (an area dominated by the first mode). On the other hand, the proposed MPA procedure which accounts for the other three transverse modes is much closer to NRHA at the end areas of the bridge. As the level of excitation increases and higher mode contributions become more significant (without substantially altering the shape of the modes) the displacement profile derived by the MPA method tends to match that obtained by the NRHA, whereas predictions form SPA become less accurate as the level of inelasticity increases. The consideration of higher modes with the proposed MPA scheme, significantly improves the accuracy of the predicted displacements, although its predictions are rather poor (but still better than those from SPA) in the areas close to the piers 5 and 8.

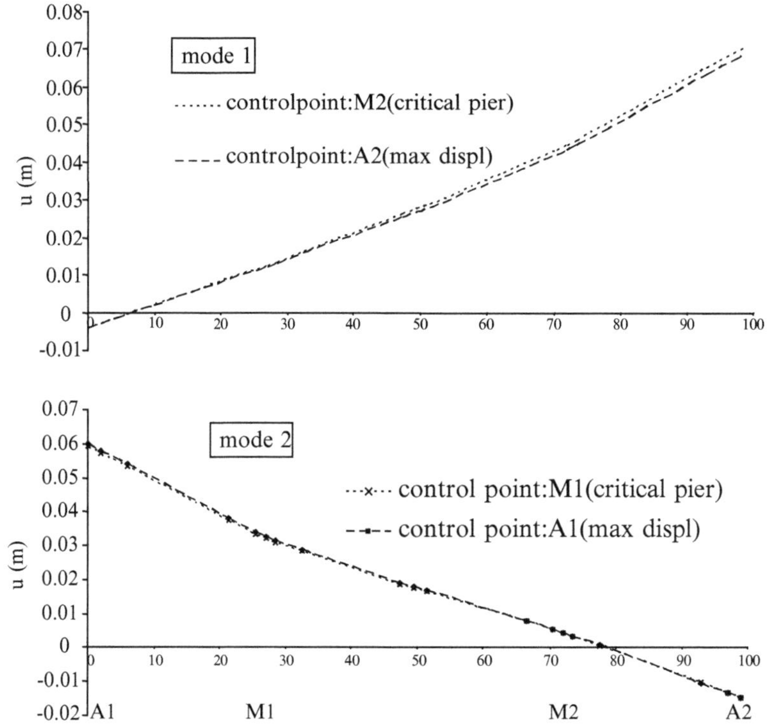

Fig. 4.13 'Modal' deck displacements derived with respect to different control points using $u_{m'}$ as target displacement according to the *improved MPA* procedure – overpass bridge ($A_g = 0.16$ g)

From Fig. 4.15, referring to the overpass bridge, it is observed that MPA predicts well (i.e. matches closely the values from the NRHA approach) the maximum transverse displacement. On the other hand, the SPA procedure underestimates the displacements of the deck at the location of the abutment A1 and the first pier of the bridge, compared to the more refined NRHA approach. This is not surprising if one notes the differences between the first two mode shapes in the transverse direction (Fig. 4.9), which are strongly affected by torsion (they contribute more than 90% of the torsional response, as well as over 90% of the transverse response of the bridge) due to the unrestrained transverse displacement at the abutments (until the 15 cm gap closes), combined with the different stiffness of the two piers caused by their different height. What is essentially achieved by the MPA is the combination of these first two modes (the 3rd transverse mode is not important in this particular bridge), each of which dominates the response in the region of the corresponding abutment. In the case of applying ground motions with twice the design earthquake intensity (also shown in Fig. 4.14), where the structure enters deeper into the inelastic range and higher mode contributions become more significant (without substantial alteration of the mode shapes) it is noted that the displacement profile derived

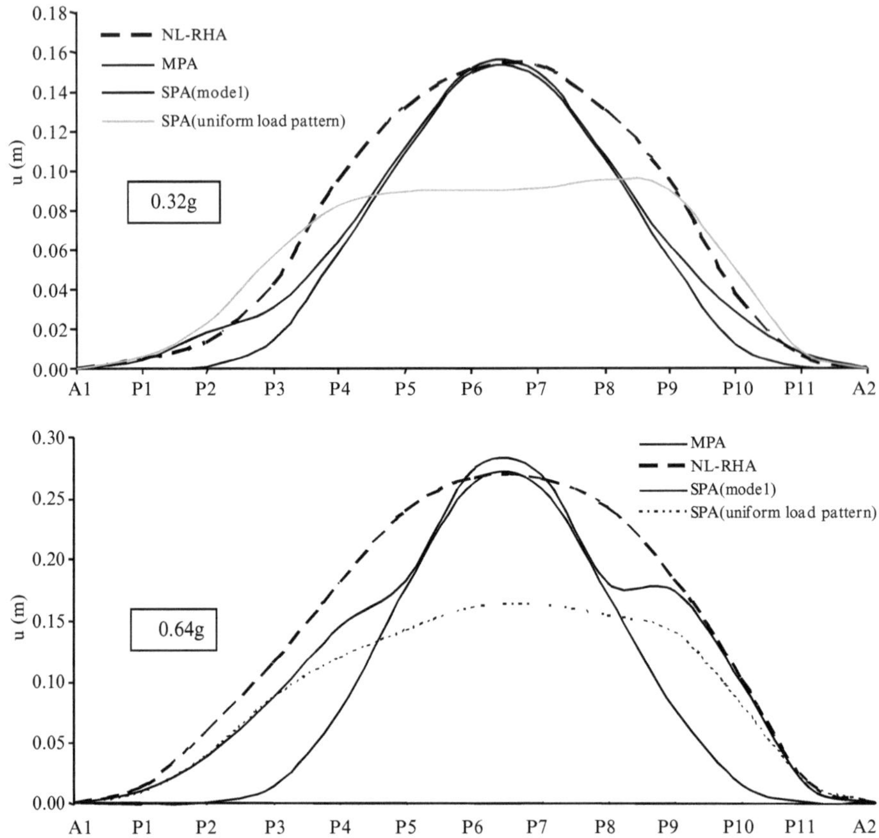

Fig. 4.14 Response to the design earthquake (0.32 g) and to twice the design earthquake, calculated from *SPA*, *MPA* and *NRHA*: deck displacements of Krystallopigi bridge

by the MPA method tends to match that obtained by the NRHA, whereas SPA's predictions remain poor. Note that, regardless of earthquake intensity, the uniform loading pattern (also shown in Figs. 4.14,–4.16) fails to capture the increased displacements towards the abutments; nevertheless its overall prediction of the displacement profile could be deemed better than that resulting from using one single modal load pattern.

From Fig. 4.16 referring to the G11 bridge, it is observed that SPA predicts very well (i.e. matches closely the values from the NRHA approach) the maximum transverse displacement, while the improvement of the displacements derived by MPA procedure is not significant even for increasing levels of earthquake excitation. This implies that the SPA method works reasonably well when applied to bridges of regular configuration, where the higher mode contribution is not significant.

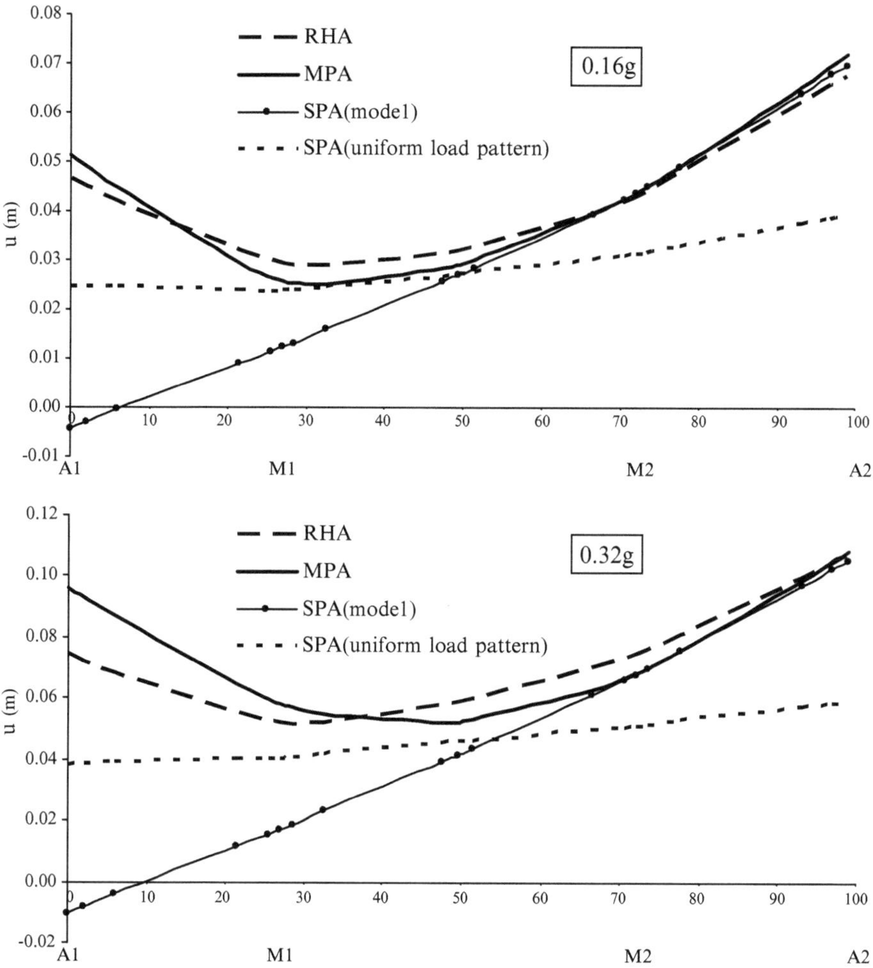

Fig. 4.15 Response to the design earthquake ($A_g = 0.16$ g) and to twice the design earthquake ($A_g = 0.32$ g) calculated from *SPA*, *MPA* and *NRHA*: deck displacements of the overpass bridge

4.3.1.4 Derivation of Multi-modal Pushover Curves

Multimodal pushover curves were derived as described in Sect. 3.5.1, and compared with dynamic pushover curves (from NRHA). The pushover curves derived using different procedures are shown in Figs. 4.17 and 4.18. For the 'dynamic' pushover curve, the base shear of the structure, as well as the transverse deck displacement at the control point, were extracted from the database of the NRHA results for each intensity level. The three combinations of base shear and maximum displacement of the monitoring point, described in Sect. 3.5.1, were used in order to derive "dynamic" pushover curves.

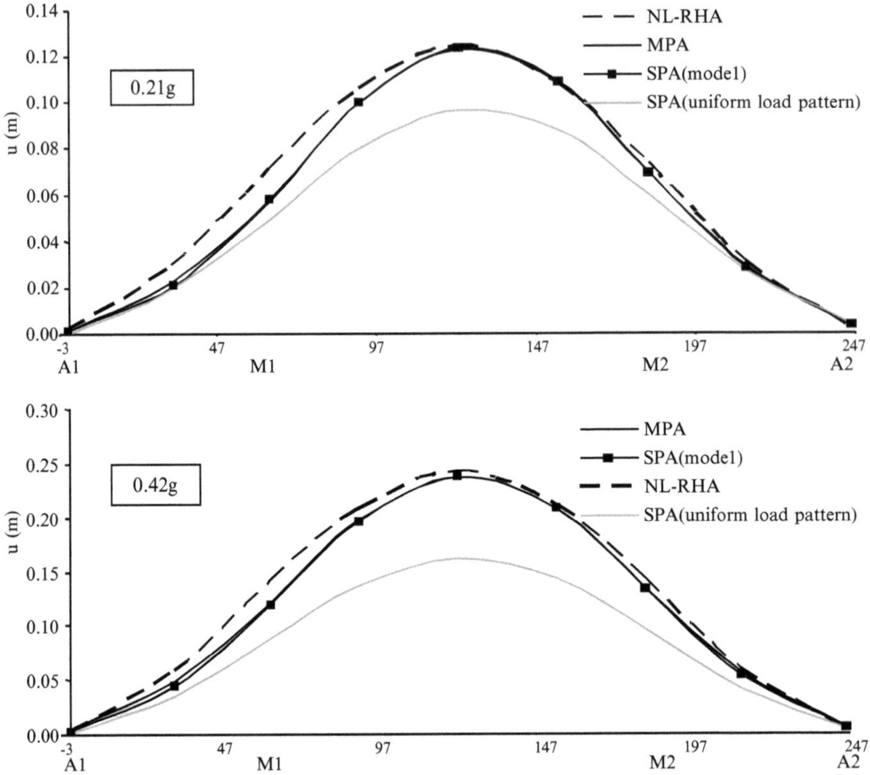

Fig. 4.16 Response to the design earthquake (0.21 g) and to twice the design earthquake (0.42 g) calculated from *SPA*, *MPA* and *NRHA*: deck displacements of the G11 bridge

In the diagrams shown in Figs. 4.17 and 4.18, the point of maximum deck displacement was taken as the control point (for Krystallopigi bridge it is located at the mass centre of the deck, while for the overpass bridge it is at the location of the abutment A2, where displacement is maximum in SPA). It is clear from the figures that the multimodal pushover curve reasonably matches the 'dynamic' one derived from the more rigorous NRHA for both structures, while the pushover curve based on SPA method is less accurate. In particular, in the case of Krystallopigi bridge (Fig. 4.17), it is observed that multimodal pushover curve is almost identical to the physically meaningful 'dynamic' pushover curves (i.e. the ones based on simultaneous quantities), while the pushover curve based on standard pushover analysis slightly underestimated the total base shear of the structure. In the case of the overpass bridge, it is clear from Fig. 4.18 that the multimodal pushover curve reasonably matches the 'dynamic' one derived from the more rigorous NRHA. On the other hand, the pushover curve based on a single modal pattern (SPA), strongly underestimated the total base shear (and the member shears) of the studied bridge.

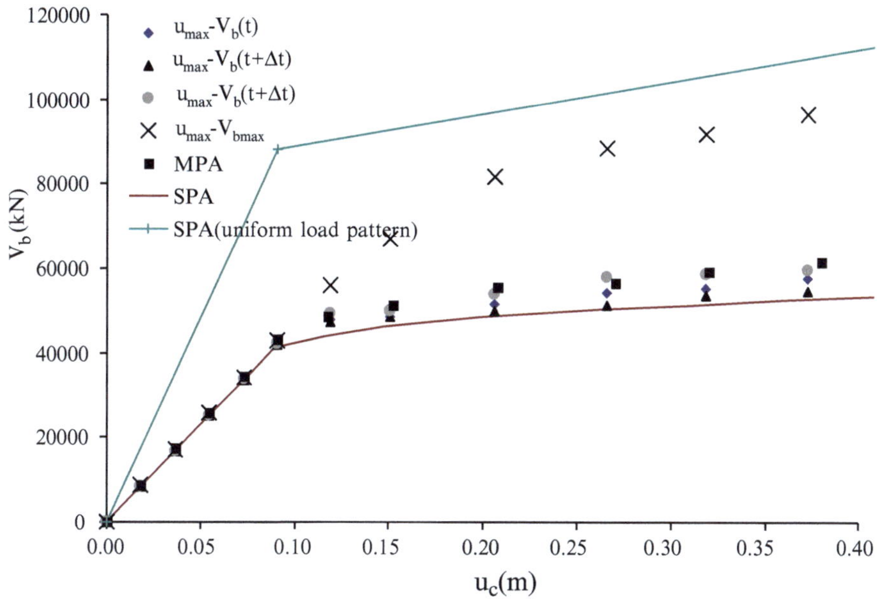

Fig. 4.17 Pushover curves derived with respect to the deck mass centre derived from *SPA* (first mode), *SPA* (uniform load), improved *MPA*, and *NRHA* for Krystallopigi bridge

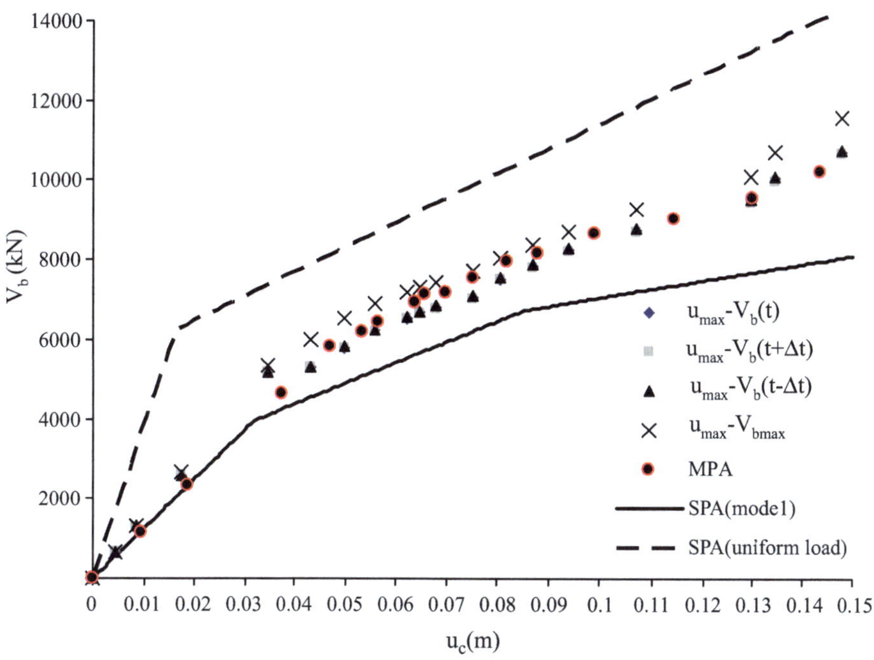

Fig. 4.18 Pushover curves derived with respect to the deck displacement at the location of the right abutment derived from *SPA* (first mode), *SPA* (uniform load), improved *MPA*, and *NRHA* for the overpass bridge

The pushover curve based on SPA using a uniform load pattern overestimated the total base shear (and the stiffness) of Krystallopigi bridge and the overpass bridge. It is notable, however, that the two SPA curves based on the load patterns recommended in modern codes (CEN 2005; ASCE/SEI 2007) envelope the true response of both bridges (the first-mode pattern provides a lower bound and the uniform pattern provides an upper bound). It is noted that the pushover curve based on SPA (load pattern proportional to the fundamental mode) is more accurate in the case of Krystallopigi bridge than for the overpass bridge, taking the dynamic pushover curves as the benchmark. This is expected from the discussion presented in the previous section where it was shown that the first transverse mode was sufficient for describing the response of the central part of the Krystallopigi bridge (to which the multimodal pushover curve refers), whereas the overpass bridge is strongly affected by torsion, and the first mode shape does not represent a realistic pattern of deck displacements of the structure.

4.3.1.5 Concluding Remarks

On the basis of the results obtained for the studied bridges, the improved MPA procedure presented in Sect. 3.5.1 appears to be a promising approach that yields more accurate results compared to the 'standard' pushover, without requiring the high computational cost of the NRHA, or of other procedures involving multiple eigenvalue analyses of the structure to define improved loading patterns in the inelastic range. The present study confirmed findings from previous studies which have indicated that SPA generally works reasonably well when applied to bridges of regular configuration (as opposed to irregular ones, such as those affected by torsion).

The multimodal pushover curves were found to reasonably match the dynamic pushover curves derived from the more rigorous NRHA. On the other hand, standard pushover curves based on single modal patterns, underestimated the total base shear (and stiffness) of the studied bridge. On the positive side, standard pushover curves based on first-mode and uniform load patterns, were found to represent a lower and an upper bound, respectively, of the more rigorous 'dynamic' pushover curves; hence they could be used whenever simplicity is prioritised higher than accuracy.

4.3.2 Case Study 2: Pushover and Dynamic Response History Analyses of Bridges

4.3.2.1 Overview of the Pushover Methods Used in This Case-Study

In this case-study three pushover methods were used: (1) the N2 method (single mode non-adaptive), (2) the MPA (multimode non-adaptive), (3) the IRSA (multi-mode adaptive). The first two methods were used in a slightly different way than the originally proposed one (see Chap. 3).

The N2 method (Fajfar and Fischinger 1987; Fajfar et al. 1997) was used in a different way than that included in the Eurocode 8 standard (CEN 2004 and CEN 2005). The most important differences are: (1) instead of the centre of the mass, the position of the maximum displacement was chosen as the monitoring point (in similar way as in Kappos and Paraskeva 2008; Paraskeva et al. 2006), (2) in addition to the distributions proposed in the standard, parabolic distribution of the inertial forces was also considered in bridges pinned at the abutments, (3) in this type of bridges bilinear instead of elasto-plastic idealization of the pushover curve was performed, (4) the optional iterative procedure (CEN 2005) for calculation of the target displacement was always employed. The basis for these changes can be found in Isaković and Fischinger (2006).

The MPA was mostly used as it was originally proposed by Chopra and Goel (2002). The only exception is the choice of the monitoring point, for which the maximum displacement point was used.

The IRSA was used as it was proposed by the author of the method (Aydinoglu 2004). All analyzed methods were tested comparing them with nonlinear response history analysis (NRHA) and experimental results.

4.3.2.2 Description of Studied Bridges

Analyzed structures were divided into two groups regarding their length: (1) short and (2) long bridges.

Short Bridges

Short bridges were four span single-column bent viaducts (see Fig. 4.19), with typical span length of 50 m. The heights of the individual columns were varied as multiples of a reference height of 7 m ($h_c = 7$ m, 2×7 m, 3×7 m). The label of any particular viaduct is Vijk, where i, j, and k denote the heights of the first, second and third columns, respectively, expressed in multiples of the reference height of 7 m. The analyzed structures are summarized in Fig. 4.19. Beside these viaducts, an experimentally tested large-scale model (Johnson et al. 2006) of a two-column bent bridge was also considered in the study (see Fig. 4.20 and Sect. 4.4.1).

The analyzed structures were divided into groups of: the regular (V232), the irregular (V123) and very irregular (V213, experimentally tested bridge) structures. The structure was considered as irregular if the influence of the higher modes to the response was considerable or the predominant mode was considerably changing with the increased seismic intensity.

All viaducts presented in Fig. 4.19 were analyzed for two seismic intensity levels; peak ground acceleration (PGA) was 0.35 and 0.7 g. In the NRHA three

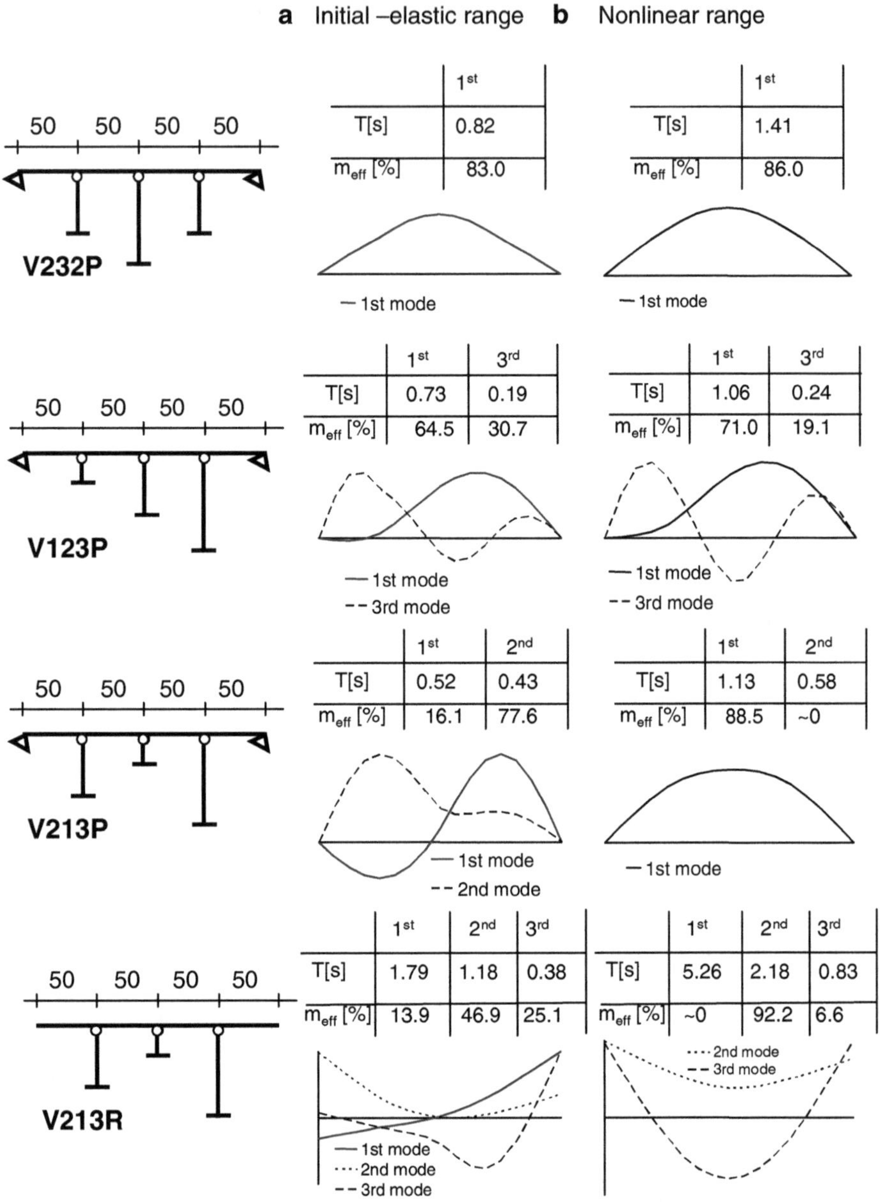

Fig. 4.19 Main dynamic properties of the analyzed bridges: (**a**) in the elastic range, and (**b**) after yielding of all columns

artificially generated accelerograms were used. They were generated based on the Eurocode 8 (CEN 1994) spectrum for soil type B. The same spectrum was used as the basis for all pushover analyses. The experimentally tested bridge was analyzed for seven different seismic intensities, the PGA varying from 0.08 to 1.65 g.

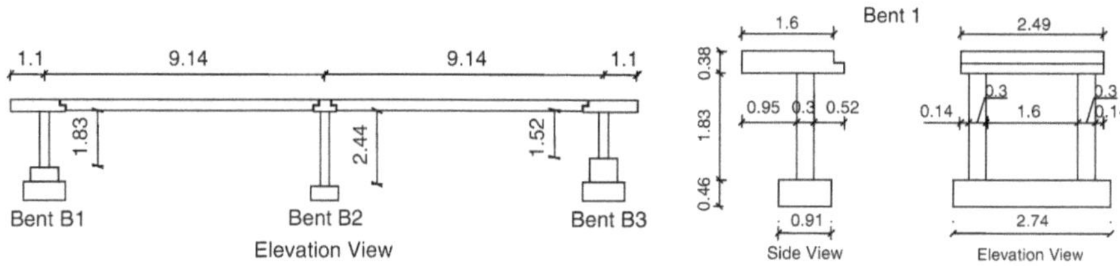

Fig. 4.20 Scheme of the experimentally tested bridge

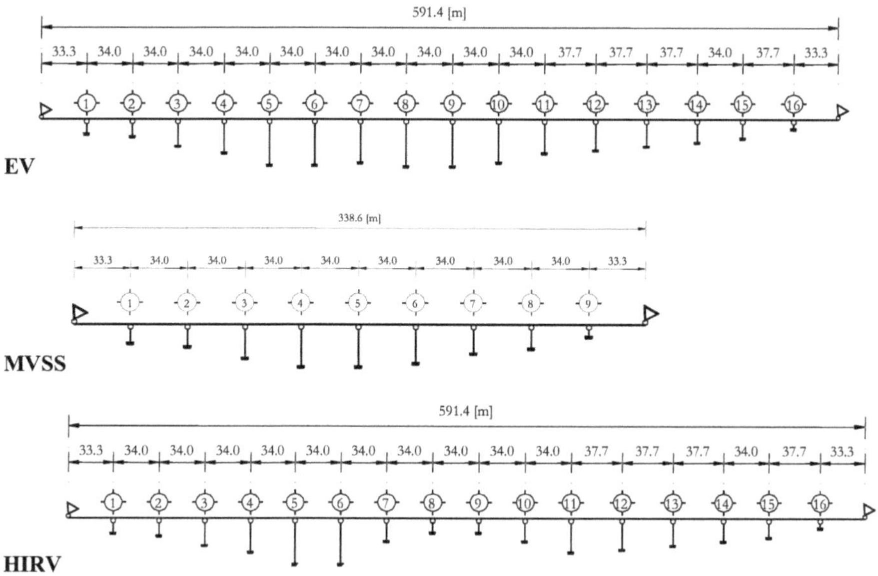

Fig. 4.21 Schemes of the analyzed long viaducts

Long Bridges

Four types of long bridges were analyzed.

1. The first structure is an existing viaduct (denoted as EV) presented in Fig. 4.21, which is 591 m long and supported by 16 single column bents. Maximum and minimum column height was 6.5 and 34.5 m, respectively. The columns have a hollow box cross-section with a longitudinal reinforcement ratio of 1.2%. The superstructure consists of four I girders, which are connected together by means of a continuous deck slab. In the elastic range (corresponding to low intensity earthquakes) the response of this viaduct was influenced by four modes (1st, 3rd, 5th, 7th). The most important were the first and third mode, which were both translational (Isaković et al. 2008). The effective modal mass m_{eff} of these two nodes amounted to 52.6% and 23.2%, for the first and the third mode, respectively.

2. A second structure, denoted by MVCSP (Modified Viaduct with the increased Cross-Sectional Properties of the original superstructure), had a stiffer superstructure than the original viaduct. To make the superstructure stiffer, the area and moment of inertia of the superstructure were increased three times, assuming that the cross section was a box girder, with a greater height than that of the original superstructure. The supporting bents were kept the same as in the original viaduct. In the elastic range the most important modes were the first and the third, with effective modal masses of 60% and 21.7% of the total mass, respectively.
3. A third structure denoted as MVSS (Modified Viaduct with Shortened original Superstructure) was defined by shortening the original structure to 339 m (Fig. 4.21). The cross-sectional properties of the original superstructure were doubled. The number of supporting bents was reduced to nine. Bents 1–4, 11–12 and 14–16 from the original viaduct were used to support the modified structure. In the elastic range the influence of the higher modes was considerable. The effective modal masses were equal to 67.0% and 20.8% of the total mass, for the first and third modes, respectively. When all columns yielded, the response was governed almost by only one mode (the effective modal mass corresponding to the first mode increased to 82.7% of the total mass).
4. A fourth structure is Highly IRregular Viaduct (HIRV) with short central columns. In the elastic range the response of the HIRV viaduct was predominantly influenced by 1st, 2nd and 5th modes. The corresponding effective masses were 24.5%, 40.9% and 26.2% for the 1st, 2nd and 5th mode, respectively. The first mode was torsional (Isaković et al. 2008).

The EV and the HIRV viaducts were analyzed for three load levels. Peak ground accelerations (PGA) of 0.125, 0.25 and 0.5 g were considered. The response of the viaducts when subjected to the weakest load was elastic. When moderate load was applied, the yield level was reached. When the structure was subjected to the strongest earthquake intensity of PGA = 0.5 g, plastic hinges occurred in all bents. The modified structures MVCSP and MVSS were subjected only to the seismic load, corresponding to a PGA of 0.5 g.

In the inelastic response-history analysis (IRHA) 100 artificially generated accelerograms were used. They were generated based on the Eurocode 8 (CEN 2004) elastic spectrum type 1 for soil type A. The same spectrum was used as the basis for all pushover analyses. More details about the analyzed bridges can be found elsewhere (Isakovic et al. 2003; Isakovic and Fischinger 2006; Isakovic et al. 2008).

4.3.2.3 Single Mode Method – The N2 Method

The two main assumptions of the N2 method are that the response of the structure is influenced by one predominant mode and that this mode does not considerably change when the seismic intensity is changed. The limitations of the method are, therefore, mostly related to these two assumptions.

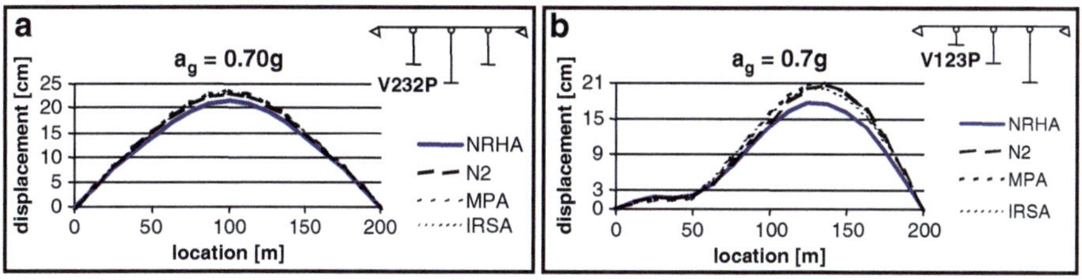

Fig. 4.22 Displacements of the superstructure in bridges *V232P* and *V123P*

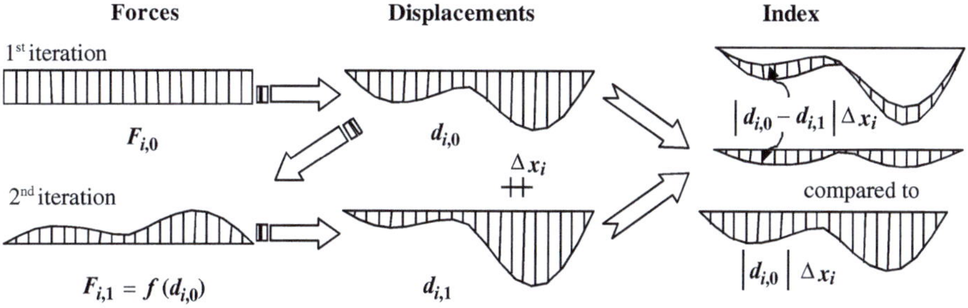

Fig. 4.23 Calculation of the index defining the applicability of the N2 method

The accuracy of the N2 method is good when the response is influenced by one predominant mode with effective mass of at least 80% of the total mass of the structure. Thus the method can be mostly used for the analysis of regular and slightly irregular structures, where the influence of the higher modes is not important. Two examples are presented in Fig. 4.22, where the response of the viaducts V232P and V123P is presented.

The regularity of a bridge and the importance of the higher modes depend on the ratio of the stiffnesses of the columns to that of the superstructure. The greater the stiffness of the superstructure is, and the smaller is that of the columns, the more regular the viaduct is, and the less important the higher modes.

Beside the previously described criterion, which could be used to test the influence of the higher modes, an additional test should be made in order to check if the predominant mode can considerably change. For this purpose an index, as proposed in Isaković et al. (2003), could be used. It is based on the normalized displacements (normalized to the maximum displacement) of the superstructure and it represents the relative difference between the areas bounded by the normalized displacement lines of the two consecutive iterations (see Fig. 4.23) of the N2 method. The procedure of index calculation is presented schematically in Fig. 4.23.

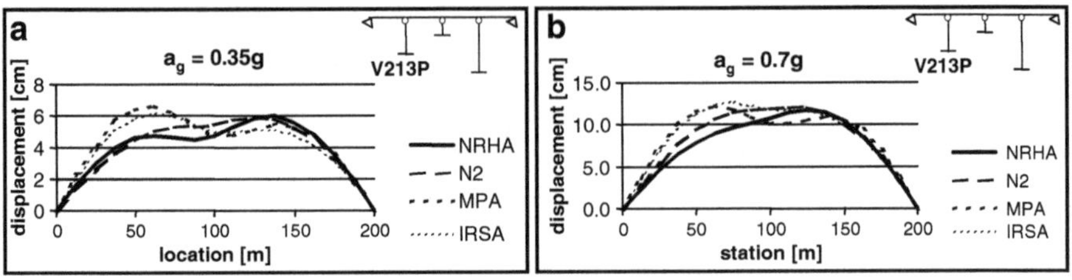

Fig. 4.24 Displacements of the superstructure in bridge *V213P*

When displacements are calculated at the equidistant nodes, the index can be expressed as:

(a) Viaducts pinned at the abutments

$$index = \frac{\sum |d_{i,0} - d_{i,1}| \Delta x_i}{\sum |d_{i,r}| \Delta x_i} \cdot 100 = \frac{\sum_{i=1}^{n-1} |d_{i,0} - d_{i,1}|}{\sum_{i=1}^{n-1} |d_{i,r}|} \cdot 100\,[\%] \tag{4.1}$$

(b) Viaducts with roller supports above the abutments

$$index = \frac{\sum |d_{i,0} - d_{i,1}| \Delta x_i}{\sum |d_{i,r}| \Delta x_i} \cdot 100$$
$$= \frac{\frac{1}{2}|d_{0,0} - d_{0,1}| + \sum_{i=1}^{n-1} |d_{i,0} - d_{i,1}| + \frac{1}{2}|d_{n,0} - d_{n,1}|}{\frac{1}{2}|d_{0,r}| + \sum_{i=1}^{n-1} |d_{i,r}| + \frac{1}{2}|d_{n,r}|} \cdot 100\,[\%] \tag{4.2}$$

The greater the value of the index is, the more irregular is the response, and consequently the more irregular is the structure. Based on the previous evaluations of the index (Isaković et al. 2003) it can be concluded that N2 method can be used in all cases where this index does not exceed the value of 5%.

In irregular bridges, the N2 method is less accurate. However, in some cases its accuracy can improve when the load level is increased. In short bridges, supported by few bents, plastic hinges usually develop in all columns at a certain load level. Their stiffness is consequently reduced and the response is mainly governed by the superstructure. When the superstructure is relatively short, its stiffness is usually large enough to make the influence of the higher modes negligible. Results of the N2 method and the NRHA then usually coincide quite well. At lower seismic intensities the matching of these two methods is usually poorer. Since in this case the response is elastic or close to elastic, the influence of higher modes is considerable. Consequently, the N2 method is less accurate.

The influence of seismic intensity on the accuracy of the N2 method in irregular viaducts is illustrated in the example of the viaduct V213P (Fig. 4.24). The match

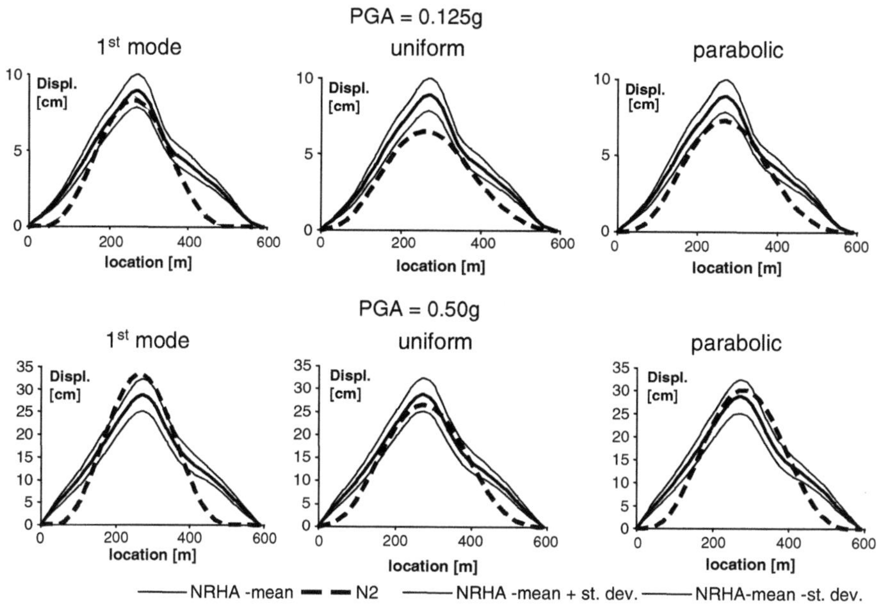

Fig. 4.25 Response of the viaduct *EV*

between displacement envelopes obtained by the N2 method and by NRHA was better in the case of the strong seismic load (Fig. 4.24b) than in the case of the weaker load (Fig. 4.24a), when the response was practically elastic.

The accuracy of the N2 method decreases with the increase of the bridge length. In long viaducts the stiffness of the typical superstructure is relatively low due to its large total length. Consequently, higher modes significantly influence the response, regardless of the seismic intensity (or the strength of the bents). An example is presented in Fig. 4.25, where the response of the EV bridge is analysed. In the N2 method it was evaluated for three different distributions of inertial forces. However, in all cases including the envelope of the results, the N2 method differed from the NRHA. When the length of the bridge was reduced, the results agreed much better with the NRHA (Isaković et al. 2008).

It is not only the higher modes that influence the accuracy of the N2 method, but also the changes of the predominant mode. When the latter considerably changes by the seismic intensity, the N2 method is less accurate. An example is presented in Fig. 4.26, where the response of the experimentally tested bridge is presented for two different seismic intensities. The measured maximum displacements of the superstructure are compared with the values, calculated by the N2 method. For the lower intensity levels the agreement is good. When the intensity of the load was increased, substantial rotations in the horizontal plane of the superstructure were observed, considerably changing the predominant mode. Consequently, the results of the N2 method were less accurate for higher intensity levels. More details can be found in Sect. 4.4.1 and elsewhere (Isaković and Fischinger 2008).

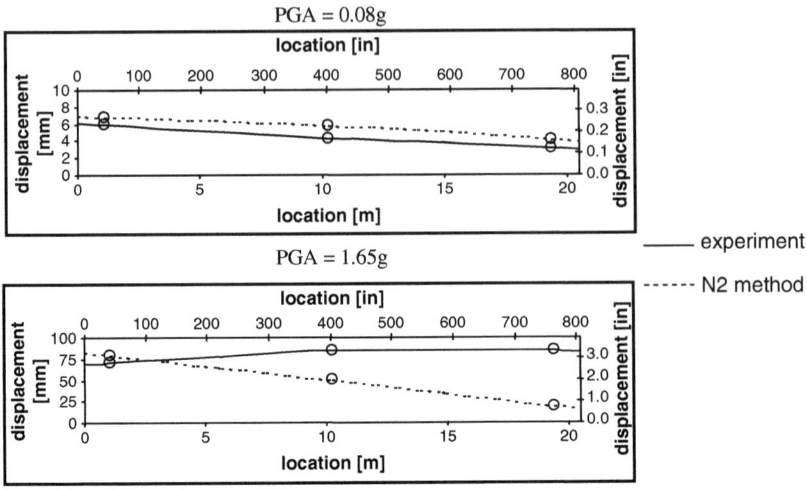

Fig. 4.26 Displacements of the experimentally tested bridge

The accuracy of the N2 method depends on the assumed distribution of inertial forces. An example is presented in Fig. 4.25. The distribution proportional to the first mode shape resulted in quite good estimation of the superstructure displacements in the central part of the bridge, whereas these displacements were underestimated in the regions close to the abutments. The uniform distribution improved the displacements in regions close to the abutments; however displacements at the central part were underestimated. Displacements corresponding to the parabolic distribution were in-between those from uniform distribution and that proportional to the first mode. When the envelope of displacements corresponding to these two distributions was taken into account (as proposed in Eurocode 8) the prediction was better, however the displacements near the abutments were still somewhat underestimated. A more extensive explanation about the choice of the distribution of the inertial forces can be found in Isaković and Fischinger (2006).

One of the crucial steps in the application of the N2 method is the idealization of the pushover curve. It can considerably influence the value of maximum displacements. The elastoplastic idealization, originally proposed for buildings, is not a suitable solution for bridges pinned at the abutments. In such bridges the bilinear approximation is more appropriate. The elastoplastic idealization in general results in overestimated maximum (target) displacements (Isaković and Fischinger 2006).

The choice of the monitoring point is the next important issue in the application of the N2 method. When the method is used for buildings, the top of the building is typically monitored. In bridges it is proposed to monitor the current position of the maximum displacement of the superstructure, taking into account the fact that in some cases this position can vary considerably as a function of the load intensity (Isaković and Fischinger 2006).

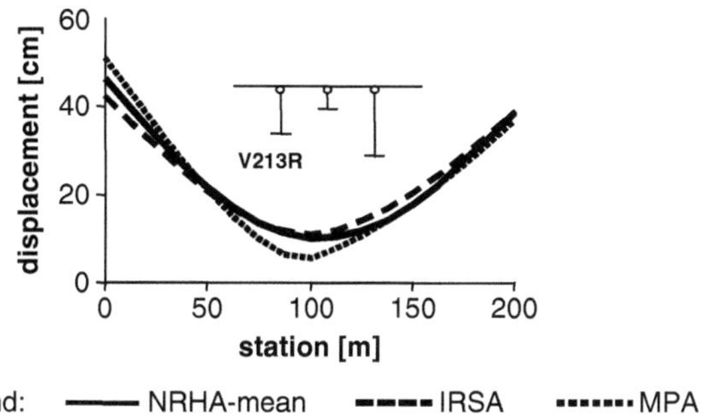

Fig. 4.27 Response of the *V213R*

4.3.2.4 Multimode Methods – The MPA and the IRSA Method

In regular structures there is no real need to use multi-mode pushover methods, since single mode pushover methods provide sufficiently accurate results. In the majority of irregular viaducts, both the investigated multi-mode methods (MPA and IRSA) correlate well with the NRHA (Isaković and Fischinger 2006; Paraskeva et al. 2006; Pinho et al. 2007).

The MPA could be less accurate for bridges subjected to strong seismic excitations, where the mode shapes change considerably, depending on the seismic intensity (see Fig. 4.27). Since the IRSA is more general, it correlates with the NRHA better than the MPA in viaducts similar to those presented in Fig. 4.27.

However, in some cases the IRSA method, too, may provide qualitatively different results from those obtained by using NRHA. Typical examples are viaducts with pinned supports at the abutments, and with very stiff central piers, e.g. the viaduct V213P (see Fig. 4.19). This is a typical example of where the dynamic properties of a viaduct change abruptly once yielding of the reinforcement occurs in the central column. In the elastic range this was a torsionally flexible structure. The response was predominantly influenced by two modes, the first of which was torsional. Subsequent to yielding of the central column, the response was predominantly influenced by only one translational mode, so that the bridge was transformed into a torsionally insensitive structure. Once the plastic hinge had occurred in the central column, the shape of the modes, as well as their order, changed significantly. It was therefore not surprising that in such cases the non-adaptive MPA could not provide accurate results. However, in the particular case the more sophisticated adaptive IRSA method was no more accurate.

A similar trend was also observed in similar long bridges. An example is presented in Fig. 4.28, where the displacements of the superstructure of the bridge HIRV are analyzed. The response of the viaduct HIRV was very complex, depending on many parameters e.g. the strength of the columns, their deformability, properties of the earthquake, etc. For some other similar configurations and combinations of these parameters the results could be substantially different (Aydinoglu and Onem 2007). Therefore it can be concluded, that the pushover methods in general should be used with care for such a type of bridges.

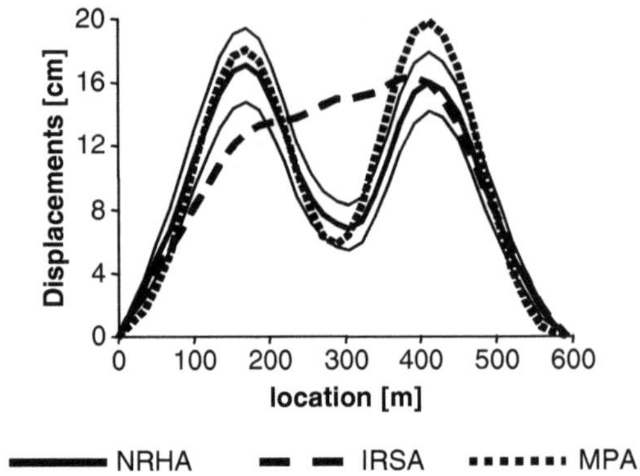

Fig. 4.28 Response of the *HIRV* viaduct

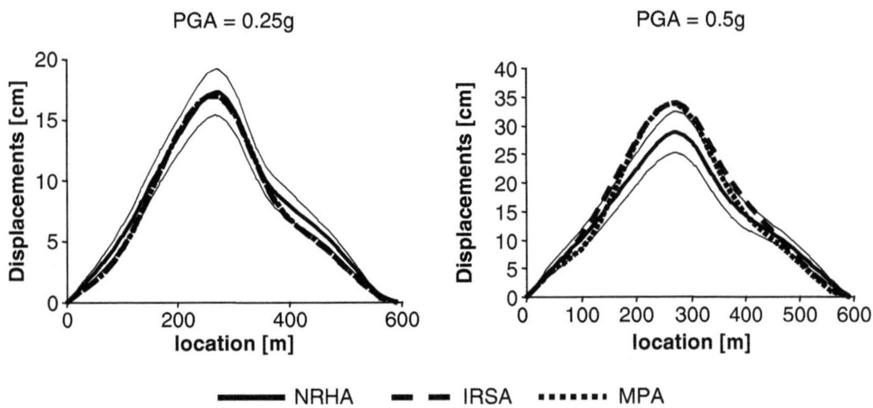

Fig. 4.29 Response of the *EV* viaduct

The use of the multimode methods is feasible in the long bridges where the single mode methods are less accurate. An example is presented in Fig. 4.29, where the displacements of viaduct EV are analyzed. As expected, the differences between MPA and IRSA were minor for the low and moderate seismic excitations (PGA = 0.125 and 0.25 g), since the corresponding response was elastic or predominantly elastic. In the elastic range both methods took into account the influence of the higher modes in the same way. The results of both methods correlated well with the results of the NRHA.

Since the important mode shapes were not significantly changed when the bridges were subjected to the strongest seismic load, the differences between MPA and IRSA were minor for this seismic intensity, too. Both methods also matched the NRHA well.

In all the moderately irregular structures the results of MPA and IRSA coincided quite well, because the mode shapes (particularly the first, most important, mode shape) did not change significantly when the seismic intensity was varied. In such cases both methods are quite efficient.

4.3.2.5 Concluding Remarks

The single mode methods such as the N2 method can be used for the analysis of bridges, where the effective modal mass of the fundamental mode is equal to or greater than 80% of the total mass and where the fundamental mode does not significantly change when the seismic intensity is changed. Changes of the mode shapes can be tested using the proposed index. In all cases where this index is less than 5% the mode shape could be considered as invariable.

In general the N2 method works well in regular and slightly irregular viaducts, where there is no significant influence of higher modes. The stiffer the superstructure is, and the more flexible are the bents, the more regular the viaduct is. Consequently, the N2 method is more accurate.

To increase the accuracy of the simplified analysis of bridges, it is recommended to take into account modifications to the N2 method listed below:

(a) More than one distribution of the inertial forces is recommended. For bridges pinned at the abutments, a parabolic distribution of the inertial forces was found to be the most suitable.
(b) The current location of the maximum displacement of the superstructure is recommended as a monitoring point during the construction of the pushover curve;
(c) Instead of the originally proposed elastoplastic idealization, a bilinear idealization of the pushover curve is recommended in bridges pinned at the abutments.
(d) It is recommended to always use an iterative procedure for calculation of the target displacements (in the Eurocode standard this is only mentioned as an option).

In short bridges pinned at the abutments and with few columns, the accuracy of single mode methods can increase with the seismic intensity. When such bridges are subjected to strong seismic excitation all the supporting columns yield. Consequently, the superstructure governs the response, the importance of the higher modes decreases, and the regularity of the bridge increases. This can improve the accuracy of the N2 method compared to the situation at lower seismic excitations.

In long viaducts (e.g. with a total length of 500 m or more) the stiffness of the typical single-spine or multi-cell box girder superstructure is relatively low due to its large total length. Consequently, the higher modes significantly influence the response, regardless of the seismic intensity (or the strength of the bents). The N2 method is less accurate. It is therefore recommended to use multi-mode pushover methods or inelastic response history for their analysis.

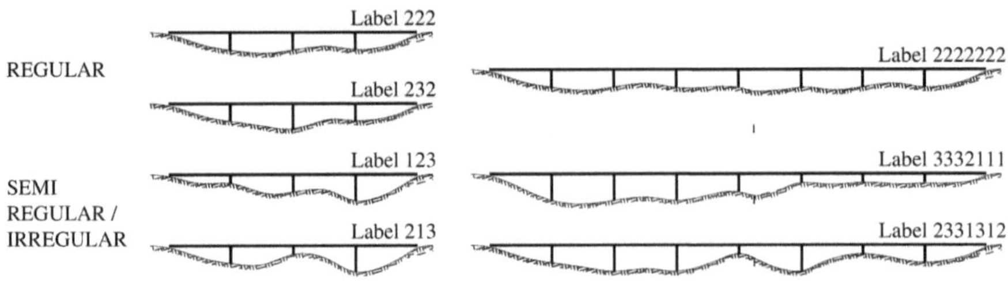

Fig. 4.30 Considered bridge configurations

In the case of the analyzed moderately irregular long viaducts (short columns near the abutments, longer columns in the central part, and gradual changes of column heights) which are frequently used in design practice, both of the investigated multi-mode methods, MPA and IRSA, performed well.

All pushover methods, including the multi-mode methods, should be used with care when torsionally sensitive structures (with short central piers) are analyzed. In some of these bridges, due to the abrupt changes in the dynamic properties of the bridge once the columns yield, the more sophisticated pushover methods can give results which are qualitatively different from those which can be obtained by using NRHA.

4.3.3 Case Study 3: Comparison of Four Different NSPs in the Assessment of Continuous Span Bridges

4.3.3.1 Parametric Study – Description

In this case-study, four commonly employed nonlinear static procedures (CSM, N2, MPA, ACSM), described in Chap. 3, are scrutinized and compared, with a view to establishing their adequacy for the seismic assessment of existing continuous span bridges.

The parametric study (Casarotti and Pinho 2007) considered two bridge lengths (viaducts with four and eight 50 m spans), with regular, irregular and semi-regular layout of the piers' height and with two types of abutments; (i) continuous deck-abutment connections supported on piles, with a bilinear behaviour (type A bridges), and (ii) deck extremities supported on linear pot bearings (type B bridges). The total number of bridges is therefore 14, as implied by Fig. 4.30, where the label numbers 1, 2 and 3 stand for pier heights of 7, 14 and 21 m, respectively. The fundamental period of vibration (see Table 4.1) ranges approximately from 0.3 to 0.5 s in short configurations and from 0.6 to 0.8 s in long ones.

The employed set of seismic excitations is defined by an ensemble of ten records selected from a suite of historical earthquakes scaled to match the 10% probability of exceedance in 50 years (475 years return period) uniform hazard spectrum for

Table 4.1 Fundamental transverse periods of vibration (seconds)

Configuration	123	213	222	232	2222222	2331312	3332111
Type A abutments	0.43	0.34	0.41	0.50	0.59	0.65	0.70
Type B abutments	0.40	0.32	0.42	0.53	0.60	0.67	0.77

Los Angeles (SAC Joint Venture 1997), which corresponds, in the current endeavour, to the intensity level 1.0. Additional intensity levels, linearly proportional to the latter by a factor of 0.5, 0.75, 1.5, 3.0 and 3.5, were also considered, thus allowing an overview on how results evolve with increasing seismic intensity. The ground motions were obtained from California earthquakes with a magnitude range of 6–7.3 recorded on firm ground at distances of 13–30 km; their significant duration (Bommer and Martinez-Pereira 1999) ranges from 5 to 25 s, whilst the PGA (for intensity 1) varies from 0.23 to 0.99 g, which effectively implies a minimum of 0.11 g (when intensity level is 0.5) and a maximum of 3.5 g (when intensity level is 3.5). The demand spectrum was defined as the median response spectrum of the ten records.

The seismic demand on the bridge models is evaluated by means of nonlinear response-history analyses (NRHA), assumed to constitute the most accurate tool to estimate the 'true' earthquake response of the structures, using the fibre-based finite elements program SeismoStruct (SeismoSoft 2006), whose accuracy in predicting the seismic response of bridge structures has been demonstrated through comparisons with experimental results derived from pseudo-dynamic tests carried out on large-scale models (Casarotti and Pinho 2006). The same software package was employed in the running of the force-based conventional pushovers (used in CSM, N2 and MPA methods) and of the displacement-based adaptive pushover analyses (Antoniou and Pinho 2004) that are required by the ACSM procedure.

Results are presented in terms of different response parameters: the estimated displacement pattern (D) and flexural moments (M) of the bridge deck at the nodes above the piers, and the shear forces at the base of the piers (V) and abutments (ABT). Then, in order to appraise the accuracy of the NSPs results obtained with the different approaches, these are normalized with respect to the median of the corresponding response quantities obtained through the incremental NRHAs; this provides an immediate indication of the bias for each of the four procedures.

Equation 4.3 shows, for a generalized parameter Δ at a given location i, how the results from the incremental dynamic analyses (IDA), run for each of the ten records considered, are first processed.

$$\hat{\Delta}_{i,IDA} = median_{j=1:10}\left[\Delta_{i,j-IDA}\right] \tag{4.3}$$

The aforementioned results' normalization consists thus in computing, for each of the parameters and for each of the considered locations, the ratio between the result coming from each NSP and the median result coming from NRHA, as illustrated in Fig. 4.31 and numerically translated into Eq. 4.4; ideally the ratio should be unitary.

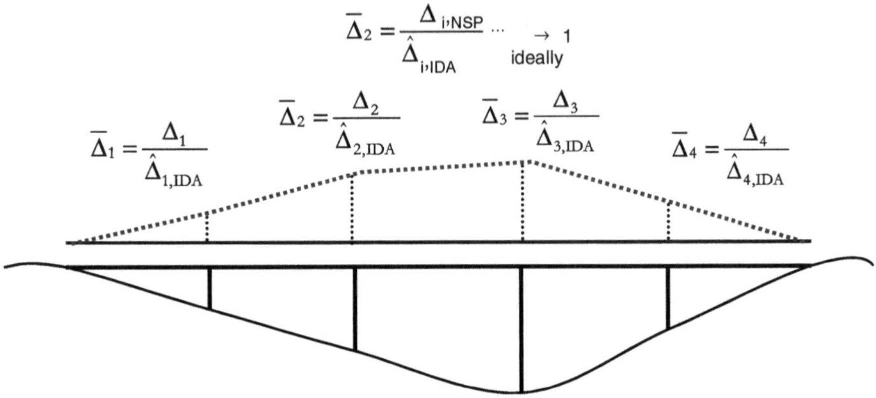

Fig. 4.31 Normalized Transverse Deformed Pattern

$$\bar{\Delta}_i = \frac{\Delta_{i,NSP}}{\hat{\Delta}_{i,IDA}} \cdots \underset{ideally}{\rightarrow} 1 \tag{4.4}$$

This normalization renders also somewhat "comparable" all deck displacements, moments and shear forces, since all normalized quantities have the same unitary target value, thus allowing in turn the definition of a so-called bridge index (Pinho et al. 2007). The bridge index (BI) is computed as the median of normalized results for the considered parameter over the m deck locations; deck displacements (BID), deck moments (BIM) or shear forces at the piers and abutments (BIV and BIABT), as shown in Eq. 4.5. The standard deviation STD measures, on the other hand, the dispersion with respect to the median, for each of the 13 procedures – Eq. 4.6.

$$BI_{\Delta,\,NSP} = \mathrm{median}_{i=1:m}\left[\bar{\Delta}_{i,NSP}\right] \tag{4.5}$$

$$\mathrm{STD}_{\Delta,\,NSP} = \left[\frac{\sum_{i=1}^{m}\left(\bar{\Delta}_{i,NSP} - BI_{NSP}\right)^2}{m-1}\right]^{0.5} \tag{4.6}$$

4.3.3.2 Parametric Study – Results Overview

In this section the results obtained from the aforementioned parametric study are scrutinized and interpreted, with a view to evaluating the accuracy of the different NSPs considered (recalled in Table 4.2). However, before passing onto a direct comparison between the four procedures, a preliminary study was carried out to identify which of the variants of the CSM, N2 and MPA methods, discussed in previous sub-sections and summarized in Table 4.2, would lead to the attainment of best results.

Table 4.2 Summary of studied nonlinear static procedures

	ACSM	CSM	N2	MPA
Pushover analysis type	Adaptive displacement-based	Conventional force-based		
Load pattern	Adaptive displacements loading	1st mode proportional loading	1st mode proportional or uniform loading	All significant modes proportional loading
Capacity curve	Base shear vs. Displacement computed from all nodes	Base shear vs. Displacement of a reference/control node, usually recommended as the centre of mass of the deck		
Demand curve	Elastic viscous damping-based reduced spectrum		Inelastic ductility-based reduced spectrum	
Considered variants	One: Preliminary study carried out elsewhere	Four: ATC 40 or FEMA 440 versions Reference node at mid-deck or at point of max displacement	Six: 1st mode, uniform or envelope loading pattern Reference node at mid-deck or at point of max displacement	Two: Reference node at mid-deck or at point of max displacement

4.3.3.3 Preliminary Evaluation

Capacity Spectrum Method

Figure 4.32 shows values of Bridge Index for each intensity level, which is the median across the entire set of bridges (see 'Intensity level results' sub-section for further details). The improvements introduced by the FEMA-440 (ATC 2005) report are clearly observed, especially in the estimations of displacement and deck moments, where results obtained using the spectrum scaling procedures suggested in FEMA-440 are much closer to unity (which means NSP estimates equal to NRHA predictions) than those obtained using the ATC-40 equations (which seem to overestimate the equivalent viscous damping, and hence the corresponding spectral reduction). The use of the location of central deck displacement as reference also (marginally) improves the results. In summary, on subsequent applications, the CSM will be employed considering its FEMA-440 version (notwithstanding the slight increase in dispersion; STD values change from 0.2–0.3 to 0.3–0.4), together with the centre of mass of the deck as reference node.

N2 Method

Figure 4.33 shows values of Bridge Index for each intensity level, considering the entire set of bridges. It is observed that whereas in the estimation of deck

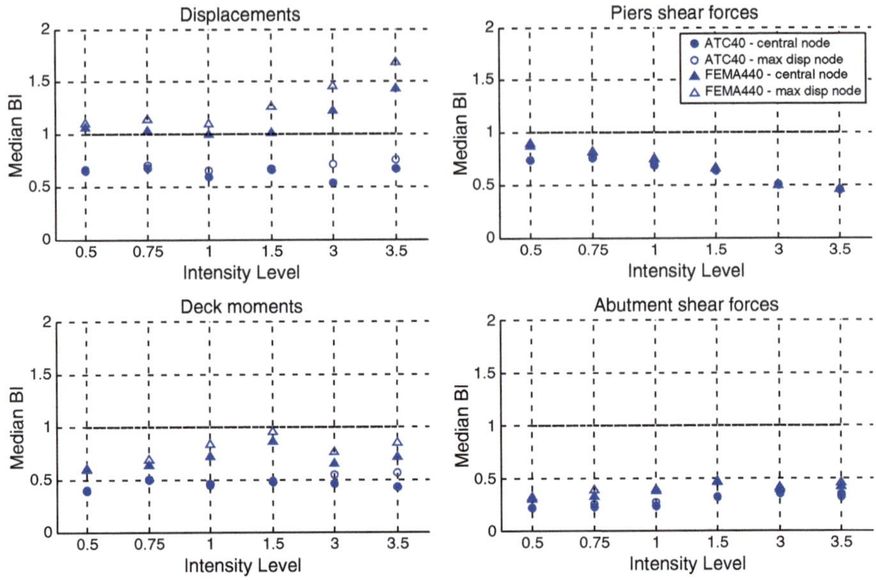

Fig. 4.32 CSM Bridge Indexes (*BI*)

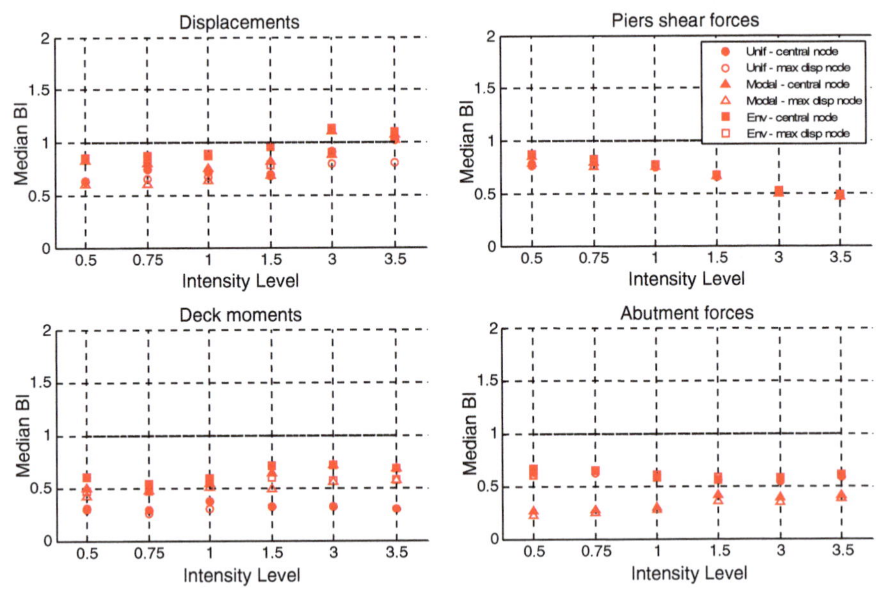

Fig. 4.33 N2 Bridge Indexes (*BI*)

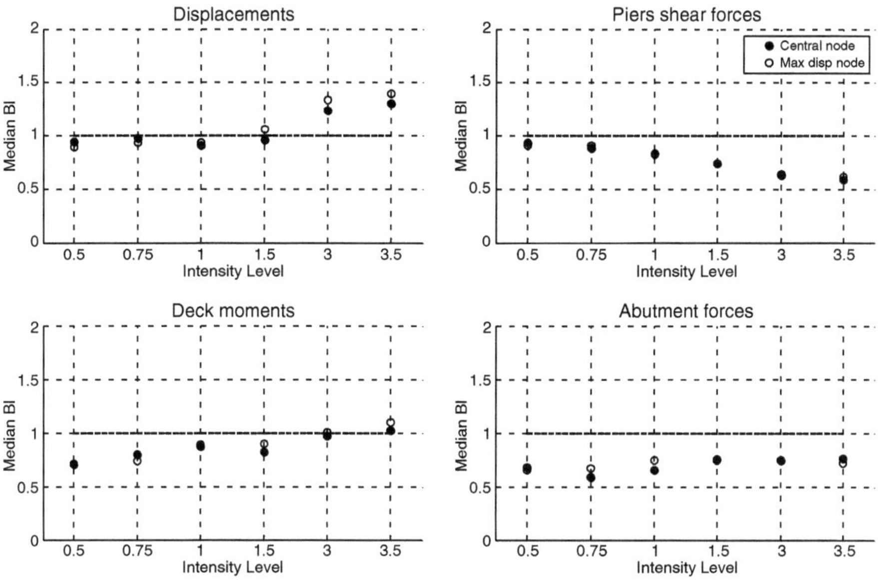

Fig. 4.34 *MPA* Bridge Indexes (*BI*)

displacements and pier shears the differences between the employment of uniform or first mode proportional load distribution does not influence the results much, when deck moments and abutment forces are instead considered the influence of pushover load shape is noticeable (and with opposite trends, which somehow explains why EC8 does not recommend the use of one loading shape over the other). The envelope shape, on the other hand, seems to somehow "contain" the positive aspects of the two EC8-recommended distributions, leading to "good" BI results for all response parameters, with relatively reduced scatter (STD varying from 0.10 to 0.30). The use of the deck's central node as reference point leads to better predictions, hence this will be adopted on subsequent applications, together with the envelope pushover loading shape.

Modal Pushover Analysis

Figure 4.34 shows values of Bridge Index for each intensity level, considering the entire set of bridges. At higher intensity levels, MPA seems to perform somewhat better when considering the maximum displacement node as reference, whilst, on the other hand, for low intensity levels, the use of the deck's centre of mass as a reference node seems to yield more consistent results. This scenario is mainly verified for displacements at pier locations and deck bending moments, given that for piers/abutments shear predictions the differences stemming from the choice of reference node are not relevant. Dispersion values proved not to be sensitive to this reference node issue, with STD values typically ranging from 0.10 to 0.30.

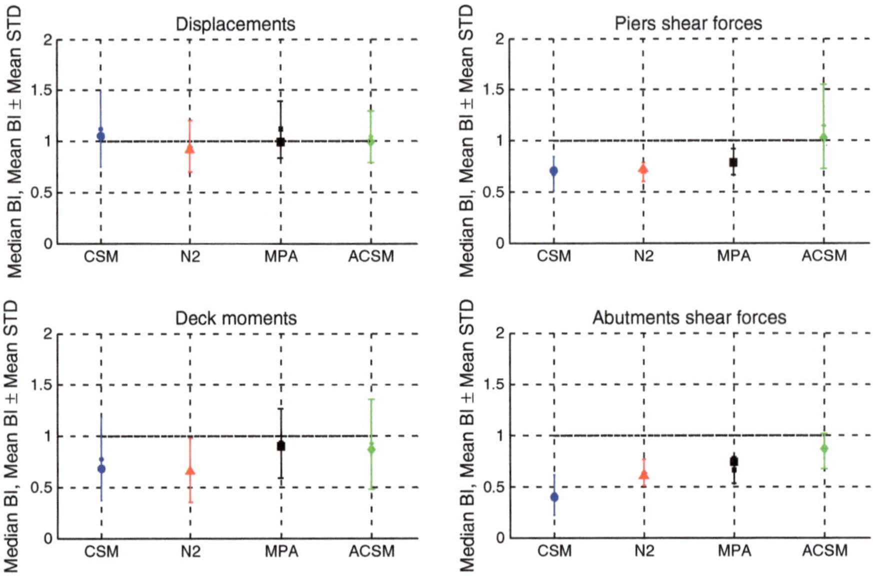

Fig. 4.35 Global Bridge Index and Standard Deviation

On subsequent applications of the method, and considering the lack of a "superior" variant, the maximum displacement reference node modality was arbitrarily adopted.

4.3.3.4 Comparative Study Results

Having identified the "optimum configuration" of all NSPs considered here, it is now possible to proceed with the parametric comparison of the four approaches, with the purpose of emphasizing relative advantages and disadvantages and, eventually, coming up with suggestions for possible preferred choices, if any. The evaluation is again carried out on the basis of Bridge Index and Standard Deviation comparison, starting from a somewhat global perspective, where the entire set of results (for all bridges and for all intensity levels) are first considered together, and then sub-structured in terms of seismic input and bridge model.

Global Results

This global results overview (Fig. 4.35, above) consists in the computation of the bridge index per NSP over all the 14 bridges and 6 intensity levels. In other words, the median bridge index over all bridge configurations and intensity levels represents the median of the single BI of every considered bridge configuration, at every intensity level. This representation of results caters for (i) comparison with

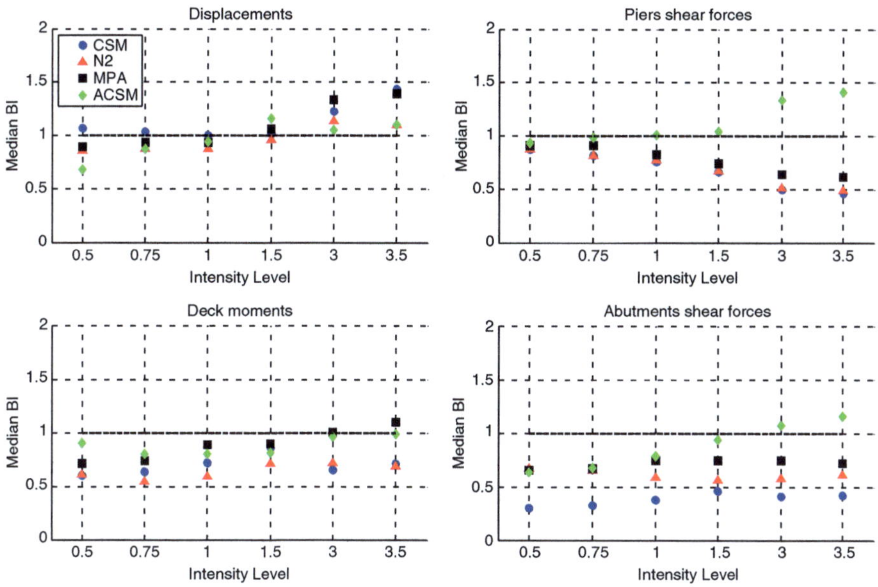

Fig. 4.36 Median Bridge Index per intensity level

response-history analyses (it is recalled that BI represents the ratio between NSP and NRHA results), (ii) relative comparison of the accuracy of the different NSPs, and (iii) appreciation of the results dispersion (plots include overall median BI for every method, in filled markers, and mean BI ± mean STD error bars).

From the observation of Fig. 4.35, it is conspicuous that all nonlinear static procedures are able to predict displacement response with relatively good accuracy, evidencing also reasonable dispersion levels. On the contrary, shear forces, either at piers or abutments, are underestimated in relatively heavy fashion, by all methods except ACSM, the latter managing to provide accurate predictions in both cases, most likely due to the employment of MMS (modified modal superposition). As for deck bending moments, MPA performs quite well, slightly underestimating NRHA results, immediately followed by ACSM, whilst, again, CSM and N2 seem to heavily underpredict such response quantities.

Regarding dispersion levels, the observed tendency is to have low scatter in shear force predictions (maximum values of 0.20) except for ACSM, which has approximately the double of that value. Higher STD values are obtained when predicting displacement and bending moment (between 0.30 and 0.40). N2 is, in most of the situations, the method with the lowest dispersion levels.

Intensity Level Results

Herein, at each intensity level, the median Bridge Index over the 14 bridge configurations is computed for each of the four NSPs (see Fig. 4.36). The results not only confirm the observations made in the previous section, but also add some insight on how these may be influenced by the intensity of the input motion. It is observed that as the latter increases the accuracy in the predictions of shear forces at the bridge piers decreases significantly. This is because pier shear forces are highly dependent on higher mode effects, which in turn become more important as the intensity of the seismic action increases (because the fundamental period elongates, hence its spectral amplification diminishes, increasing the relative importance of higher modes). N2 and CSM produce inevitably similar results, because they are both based on first mode response only, whilst MPA and ACSM, on the other hand, do consider higher modes, albeit in a different manner with respect to each other, which leads to underestimation and overestimation of forces, respectively. The latter is also visible in the variation of abutment forces, underestimated by MPA and overestimated by ACSM, which thus results conservative.

Important variations are also observed in displacement response estimates, with slight underprediction at lower intensity levels, for all NSPs, to modest overprediction at high intensity, for CSM and MPA, while N2 and ACSM seem to keep a steadier closeness to NRHA results. These differences between the four methods may be justified with the fact that major conceptual differences exist between them, such as the reference node choice, the use of an envelope of different displacement shapes for the case of N2, etc. For what concerns the dispersion of the results, this did not prove to be dependent on intensity level.

Bridge Configuration Results

Herein, for each bridge configuration, the median Bridge Index and Standard Deviation across the 6 intensity levels is plotted considering each of the four nonlinear static procedures (see Figs. 4.37 and 4.38). A bridge configuration detailed level of results enables the analysis of the influence of symmetry, regularity, length, abutments type, among other variables, on the global results previously presented and discussed.

As expected, the response predictions do appear to be very bridge-dependent, even if the observations/conclusions previously drawn still hold for the majority of configurations. The best response estimates (in terms of BIs being close to unity, and STDs being close to zero) are obtained for the regular bridge configurations (e.g. 222, 232 and 2222222), as one would expect, though good displacement estimates are also obtained for semi-regular and irregular bridges; this certainly constitutes good news for NSPs and their application to the assessment of bridge response when focusing on deformations. An overestimating trend for short bridges and an underestimating one for longer ones, can also be observed in the results, whilst the largest scatter is, in general, associated to long bridges. Finally, no relevant differences between bridges with abutments of type A (continuous deck-abutment connections) or type B (deck supported on linear pot bearings) are noticeable.

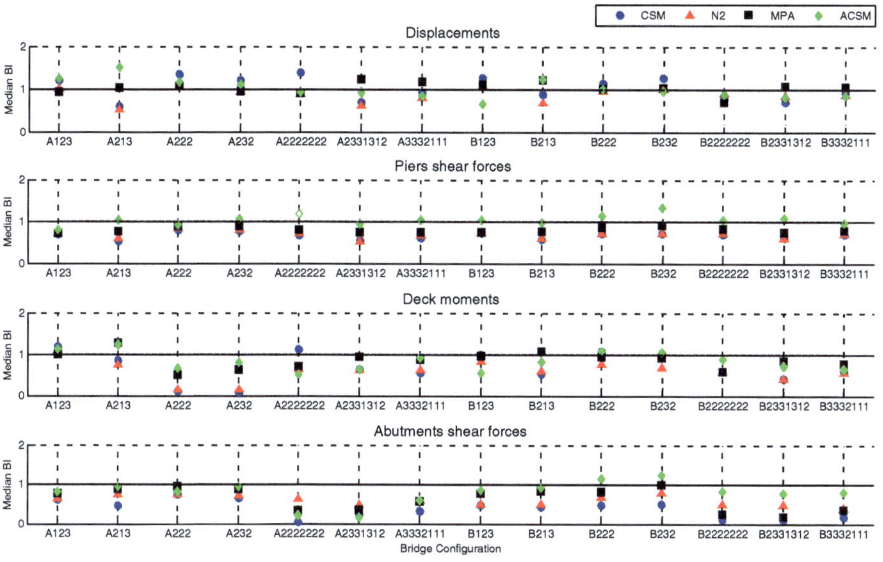

Fig. 4.37 Median Bridge Index per bridge configuration

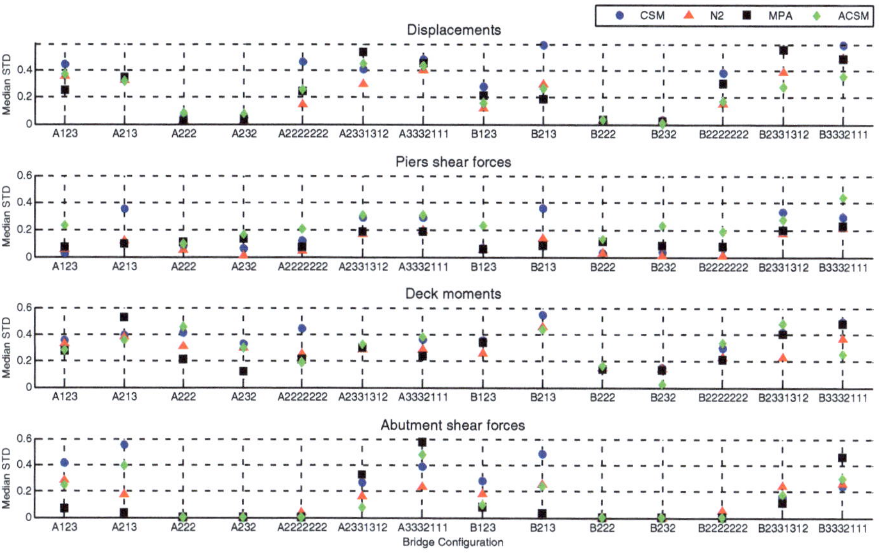

Fig. 4.38 Median Standard Deviation per bridge configuration

4.3.3.5 Concluding Remarks

The ability of four commonly used Nonlinear Static (Pushover) Procedures in predicting the structural response of bridges subjected to earthquake action has been appraised and compared; two pioneering "classical" methods (CSM and N2) were considered along with two of their more contemporary counterparts (MPA and ACSM). The evaluation was systematically carried out over a relatively large number of structural configurations, considering different response parameters and using several accelerograms scaled to a number of intensity levels. A preliminary study was also carried out with a view to establishing the optimum configuration for each of the employed procedures. The following main observations could be made:

- The Capacity Spectrum Method has clearly benefited from the improvements introduced in the FEMA-440 report, which allowed the attainment of superior predictions, with respect to those obtained using the antecedent ATC-40 formulae.
- The N2 method is favoured by the consideration of an envelope pushover load shape that bounds the two alternative load profiles suggested by EC8; first mode proportional and uniform. It is also recalled that the optional iterative procedure to optimize the equivalent bi-linearization process was employed here; this might have contributed positively to the performance of the method.
- In average all four NSPs proved to be able to predict displacement response with relatively good accuracy for all sorts of bridge configurations (regular, irregular, short, long, etc.), something that certainly does lend some reassurance with regards to the employment of such methodologies for assessing response displacements and deformations.
- The Adaptive Capacity Spectrum Method behaved in a slightly more consistent manner, given that, additionally to displacements, it managed to provide reasonable predictions for shear forces and bending moments. On the other hand, using the other procedures, shear forces were typically underestimated by at least 20–30%, the same occurring with bending moment estimates for the case of Capacity Spectrum Method and N2 Method.
- If a single NSP should be recommended over the rest, ACSM would probably be the most reliable choice, since results show that it is the only NSP that maintains fair prediction levels for all response parameters. The use of a modified version of the MPA procedure (Chopra et al. 2004), where elastic modal analysis is employed to compute forces, could perhaps improve the shear force estimates obtained with the latter. And the same could perhaps apply also to the CSM and N2 approaches. In such case then, personal preference would condition the choice of NSP to be used.

It is noted here that the case-studies in this section did not include the modified MPA procedure used in Sect. 4.3.1.

4.3.4 Case Study 4: Performance-Based Seismic Assessment of Simply-Supported Deck Bridges

4.3.4.1 Overview of the Procedure Used

In this section the seismic performances of a number of existing bridges of the Italian A16 highway are examined through Nonlinear Static Analysis (NSA). The NSA predictions are then compared with the results of Nonlinear Response History Analyses (NRHA).

The seismic performances of the selected bridges are first evaluated, through an Inverse (I) application of the Adaptive Capacity Spectrum Method (ACSM) proposed by Casarotti and Pinho (2007) and described in Sect. 3.5.3.2. For this reason, the acronym IACSM is used to identify the adopted methodology (Cardone et al. 2011). Contrary to ACSM, the IACSM is not iterative and does not require the bilinearization of the capacity curve of the equivalent SDOF model of the bridge. In the direct application of the ACSM, the main objective is to find the performance point (i.e. the damage state) of the structure under a predefined seismic intensity (i.e. a given PGA) using an over-damped demand spectrum derived by an iterative procedure for an idealized bi-linear capacity curve (ATC 1996). The main objective of the IACSM, instead, is to evaluate the seismic intensity (i.e. the PGA) of the expected ground motions, corresponding to pre-determined damage states of the structure, identified by given performance points on the capacity curve of the bridge. The inelastic deformed shape of the bridge corresponding to each damage state, therefore, is already known at the beginning of the analysis. As a consequence, the equivalent damping ratio of the bridge can be directly evaluated by properly combining the damping contributions of the single bridge components.

The capacity curves of the selected bridges have been derived through Displacement-based Adaptive Pushover (DAP) analysis (Antoniou and Pinho 2004), carried out with the FEM program Seismostruct (Seismosoft 2006), separately in the transverse and longitudinal direction of the bridge. The DAP curves were then converted into an equivalent SDOF adaptive capacity curve, according to the approach proposed by Casarotti and Pinho (2007).

Bridge performances are evaluated referring to a number of specified Damage States (DSs) of the critical members of the bridge (piers, abutments, bearing devices and joints). Each DS is identified by a given Performance Point (PP) on the DAP curve of the bridge. Herein, the DSs are related to three Performance Levels (PLs), based on the consequences in terms of damage that the attainment of each DS can produce (see Table 4.3).

For each DS, the seismic demand is evaluated by a highly-damped elastic response spectrum. This step requires the evaluation of the equivalent viscous damping of the bridge associated to that DS. To this end, the following routine has been followed (Dolce et al. 2007): (i) choose a given DS, (ii) enter the pushover database to determine the corresponding deformed shape of the bridge and the actual displacement of each structural member, (iii) evaluate the equivalent

Table 4.3 Performance levels and damage states considered in the proposed procedure

PL1 (Slight damage state)	PL2 (Moderate damage state)	PL3 (Severe damage state)
Joint closure (DS11)	Pier 50% ultimate ductility (DS22)	Steel hinges failure (DS31)
Pier yielding (DS12)	Bearings sliding displacement (DS22)	Pier ultimate ductility (DS32)
Neoprene pad failure (DS13)	Abutment active resistance (DS23)	Pier shear failure (DS33)
		Abutment passive resistance (DS34)
		Deck unseating (DS35)

damping of each structural member, (iv) combine the damping contributions with a suitable combination rule to get the equivalent viscous damping of the entire bridge.

The equivalent damping of the bearing devices has been calculated based on the well-known Jacobsen approach (Jacobsen 1930):

$$\xi_{b,j} = \frac{E_{visc} + E_{hyst} + E_{fr}}{2\pi \cdot F_{PP} \cdot d_{b,j}} \tag{4.7}$$

in which E_{visc}, E_{hyst} and E_{fr} identify the energy dissipated by the device, through its viscous, hysteretic or frictional behaviour, in a cycle of amplitude $d_{b,j}$, being $d_{b,j}$ the displacement of the device at the selected DS and F_{PP} the corresponding force level.

As far as piers are concerned, reference has been made to the relationship proposed by Kowalsky et al. (1995):

$$\xi_{p,j} = \frac{1}{\pi}\left(1 - \frac{(1-r)}{\sqrt{\mu}} - r\mu\right) \tag{4.8}$$

which relates the equivalent hysteretic damping of the pier to its displacement ductility (μ) and post-yield hardening ratio (r).

The equivalent damping of each pier-bearings system is then computed, by combining the damping values of pier and bearing devices in proportion to their individual displacements:

$$\xi_j = \frac{\xi_{b,j} \cdot d_{b,j} + \xi_{p,j} \cdot d_{p,j}}{d_{b,j} + d_{p,j}} \tag{4.9}$$

Finally, the equivalent damping ratios of the pier-bearings systems are combined to provide the equivalent damping ratio of the bridge as a whole, for the selected DS. The approach followed is to weigh the damping values of the single pier-bearings systems in proportion to the corresponding force levels:

$$\xi_{PP} = \frac{\sum_{j=1}^{n} \xi_j \cdot F_j}{\sum_{j=1}^{n} F_j} = \frac{\sum_{j=1}^{n} \xi_j \cdot F_j}{V_b} \tag{4.10}$$

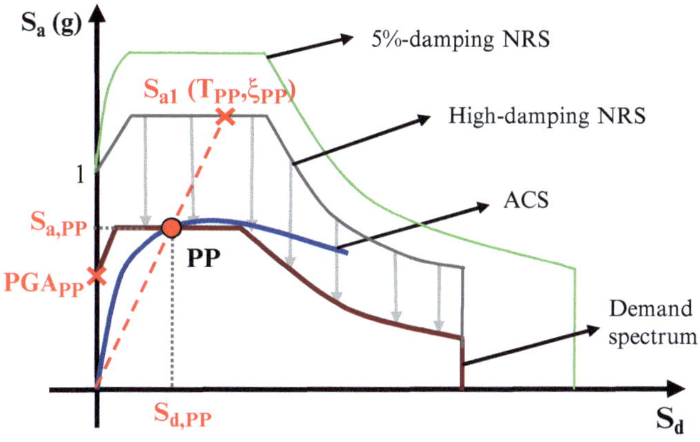

Fig. 4.39 Evaluation of the *PGA* associated to a selected performance point (*PP*)

Once the equivalent damping of the bridge has been determined, the corresponding demand spectrum can be derived from the reference 5%-damping normalized response spectrum, using a proper damping reduction factor (Cardone et al. 2008). In the proposed methodology, the damping reduction factor adopted in an old version of the Eurocode 8 (CEN 1994):

$$\eta = \sqrt{\frac{7}{(2 + \xi_{PP})}} \tag{4.11}$$

has been employed, with the limitation $\eta > 0.55$.

The PGA values associated to each DS are finally obtained by a proper translation of the high-damped Normalised Response Spectrum (NRS) to intercept the Adaptive Capacity Spectrum (ACS) in the selected PP (see Fig. 4.39).

From an analytical point of view, the PGA associated to each PP can be determined as the ratio between the acceleration of the capacity curve $S_{a,PP}$ corresponding to that PP and the normalized spectral acceleration $S_{a1}(T_{PP}, \xi_{PP})$ at the effective period of vibration T_{PP} and global equivalent damping ξ_{PP} associated to that PP :

$$\mathrm{PGA}_{PP} = \frac{S_{a,PP}}{S_{a1}(T_{PP}, \xi_{PP})} \tag{4.12}$$

The seismic performances predicted by the IACSM have been compared with the results of Nonlinear Response-History Analyses (NRHA). NRHAs have been performed with SAP2000_Nonlinear (Computers and Structures 2005), using a set of seven artificial accelerograms compatible (on average) with the EC8-Ground B response spectrum, scaled to the PGA values associated with each DS (see Eq. 4.12). The comparison has been made in terms of envelopes of bridge displacement profiles in the longitudinal and transverse direction.

4.3.4.2 Case Studies

A set of nine existing bridges of the A16 Napoli-Canosa Italian highway has been selected as case study. The selection resulted from a preliminary examination of the A16 highway, aimed at identifying bridge types and characteristics representative of the whole bridge inventory of the A16 Highway. Figure 4.40 shows the schematic layout of the examined bridges. For each of them, the bridge geometry and pier type (SS: single shaft, SF: simple frame, SW: single wall) are pointed out.

All the selected bridges are multi-span simply supported deck bridges, with span lengths of approximately 33 m. In Table 4.4 the main characteristics of each bridge structure are reported. They include: (i) pier type, (ii) bearing device type, (iii) span length, (iv) variability of pier heights (VH), (iv) longitudinal and transverse reinforcement ratios (ρ_l and ρ_s respectively) of the piers, (v) pier concrete strength and (vi) yield stress of pier reinforcement.

The variability of pier heights (VH) has been evaluated as percent difference between the average and minimum height of the piers of each span. The highest values of VH are reported in Table 4.4. The longitudinal reinforcement ratio has been computed as the total area of steel reinforcement divided by the area of the pier cross section. The transverse reinforcement ratio, instead, has been computed as the volume of transverse reinforcement divided by the volume of concrete between a single spacing.

Different types of bearing devices have been found, which basically realise three different types of pier-deck connections, i.e.: (i) fixed (SH: steel hinges), (ii) moveable (SP: steel pendulum, SR: steel rollers, S: Sliding bearing) and (iii) semi-rigid (NP: neoprene pads, DB: Dowel Bars) connections.

All the bridges taken into consideration were built between 1969 and 1972, according to the pre-1971 Italian Regulations for RC structures. As a consequence, though being built in seismic areas, they were not designed to resist any horizontal action. In addition, neither specific design criteria (capacity design) nor specific rules for seismic detailing (minimum amount of reinforcement, maximum stirrup spacing at the beam and column ends, etc.) were followed.

4.3.4.3 Modelling Assumptions

As far as modelling of the bridge is concerned, the so-called Structural Components Modelling (SCM) approach (Priestley et al. 1996) has been followed. According to the SCM approach, the bridge structure can be divided in a number of independent rigid diaphragms, modelling the bridge decks, mutually connected by means of a series of nonlinear springs, modelling bearing devices, piers and abutments (see Fig. 4.41). The deck mass is lumped in the centre of mass of each deck. A tributary mass of the pier has been lumped at the top of each pier.

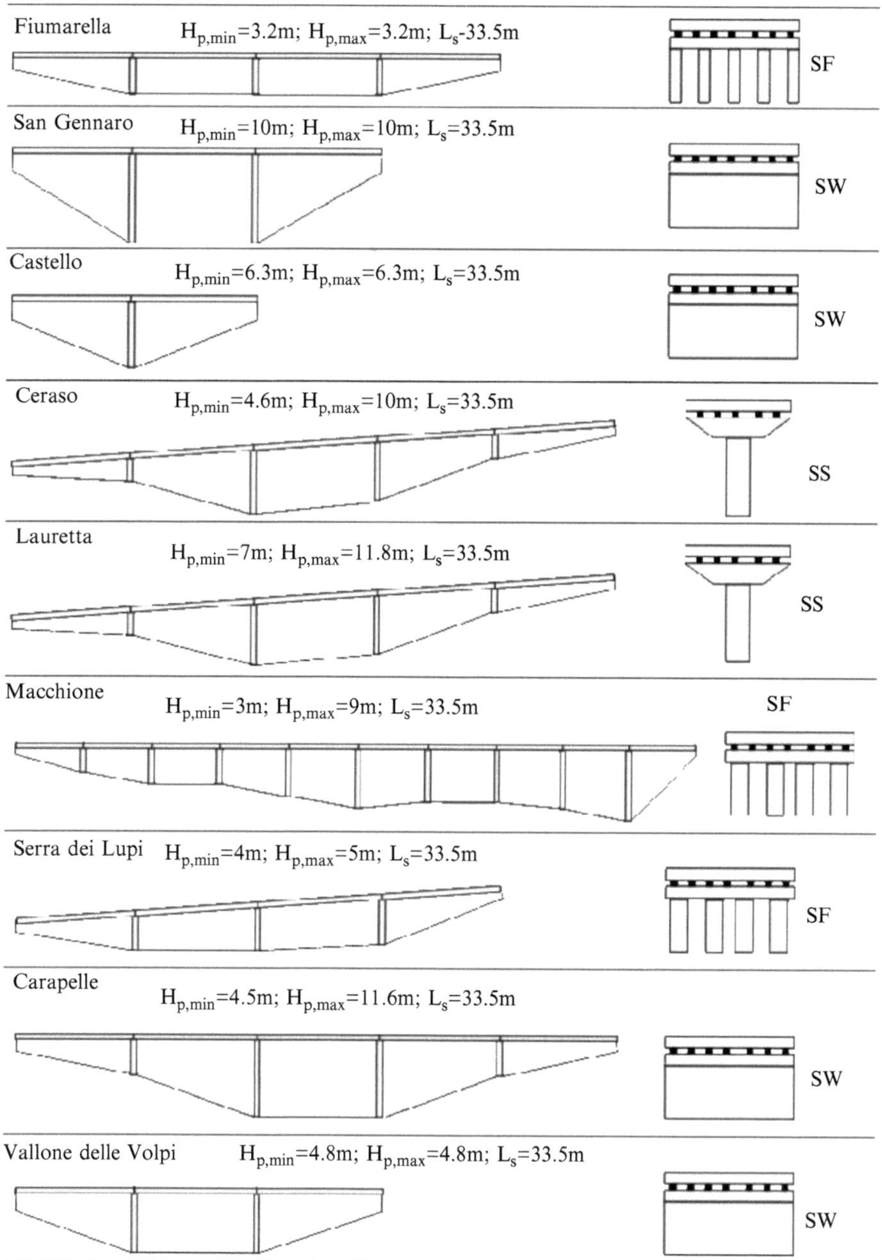

Fig. 4.40 Schematic layouts of the examined bridges

Table 4.4 Main characteristics of the analyzed bridges

Bridges	Pier types[a]	Bearing device types[b]	Span length (m)	Variability of heights (VH) (%)	Longitudinal reinforcement ratio (ρ_l) (%)	Transverse reinforcement ratio (ρ_s) (%)	Concrete strength (MPa)	Steel yielding (MPa)
Fiumarella	SF	SH(tr), SH + SP(lon)	33.50	0	0.44	0.14	35–40	270
San Gennaro	SW	NP, S	33.50	0	0.32	0.07	35–40	230
Castello	SW	NP, S	33.50	0	0.24	0.29	42.50	270
Ceraso	SS	NP	33.50	35.92	0.55	0.16	35–40	440
Lauretta	SS	NP	33.50	25.86	0.55	0.16	35–40	440
Macchione	SF	SH(tr), SH + SR(lon)	33.00	25.00	1.38	0.11	35–40	230
Serra dei Lupi	SF	DB + SH(tr), DB + SR(lon)	33.00	11.11	1.38	0.11	35–40	230
Carapelle	SW	NP, S	32.50	43.92	0.30	0.085	35–40	270
Vallone delle Volpi	SW	DB + SH(tr), DB + SR(lon)	33.50	0	0.44	0.15	35–40	440

[a] *SF* Simple Frame, *SS* Single Shaft, *SW* Single Wall

[b] *SH* Steel Hinges, *SP* Steel Pendulum, *SR* Steel Rollers, *S* Sliding bearing, *NP* Neoprene Pad, *DB* Dowel Bars

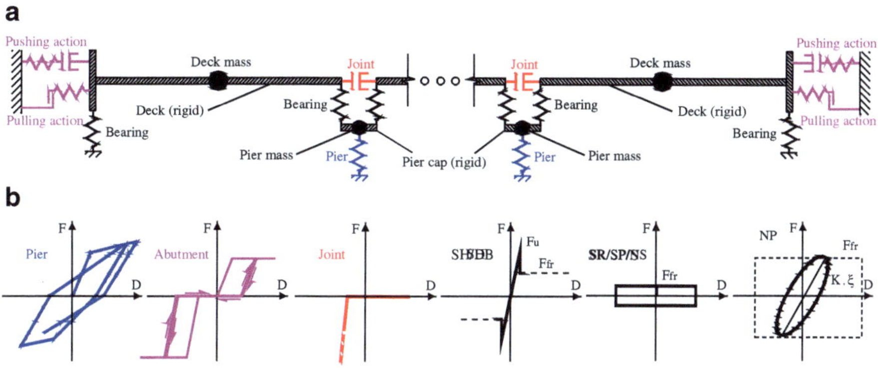

Fig. 4.41 Modelling assumptions based on the *SCM* approach: (**a**) schematic layout of the numerical model of the bridge, (**b**) phenomenological force-displacement behaviour of each bridge component

Table 4.5 shows the basic modelling assumptions made in the numerical analyses, for each bridge component, with reference to its monotonic and cyclic behaviour.

As can be seen, decks, abutments and foundations have been considered infinitely rigid and resistant, due to their low seismic vulnerability, compared to piers and bearing devices.

Possible effects due to the closure of the joints have been taken into account in the analyses. Indeed, gap closure results in pounding between adjacent decks and/or between deck and abutment, which can determine a significant redistribution of the seismic forces between piers and abutments. Joints have been modelled with compression-only translational (longitudinal direction) and rotational (transverse direction) link elements, with an initial gap assigned based on the clearance of the joint. In this study, a gap of 20–50 mm has been considered, which corresponds to the typical width of the expansion joints of the Italian bridges.

Abutments are modelled only in the longitudinal direction, with a couple of nonlinear springs, characterised by two different elastic-perfectly-plastic backbone curves (see Fig. 4.41b), modelling the pushing and pulling action of the abutment, respectively. In this study, the horizontal stiffness and ultimate strength of the abutment have been derived from a combination of design recommendations (Caltrans 2006) and experimental test results on seat-type abutments with piles (Maroney et al. 1993), as a function of the abutment back-wall dimensions and pile characteristics.

Piers have been modelled as spring elements with nonlinear force-displacement behaviour. First, a moment-curvature analysis of the critical section(s) of the pier has been performed (see Fig. 4.42a), considering concrete confinement, lap-splices and buckling effects of reinforcing steel bars (see Fig. 4.42a). The force-displacement behaviour of each pier has been then evaluated, based on simple kinematic and equilibrium relations, taking into account P-Δ effects and possible shear failure (Fig. 4.42b). The cyclic behaviour of the piers has been described by the Takeda degrading hysteretic model (see Fig. 4.41b).

Table 4.5 Basic modelling assumptions for each bridge component considered in the nonlinear static analysis (*NSA*) and in the nonlinear response-history analysis (*NRHA*)

Component	Modelling assumptions	Monotonic behaviour (NSA)	Cyclic behaviour (NRHA)
Foundations	Infinitely rigid and resistant	–	
Decks	Infinitely rigid and resistant Lumped mass	Diaphragm behaviour	
Joints	Compression-only translational (longitudinal) and rotational (transverse) springs with gap	Rigid behaviour after gap closure	
Abutments	Transverse: infinitely rigid and resistant Longitudinal: effects of piles-ground and backfill-abutment interaction considered	Elastic-Perfectly-Plastic backbone curve	Kinematic multilinear plastic model
Piers	Beam with plastic hinges at the end(s) Flexural behaviour derived from moment-curvature analysis Shear strength and P-Δ effects considered	Elastic–plastic with hardening backbone curve	Takeda degrading-stiffness hysteretic model
Bearings	Post-failure pier-deck sliding considered	Multilinear backbone curve	Viscous, hysteretic or frictional cyclic behaviour, depending on bearing type and displacement amplitude

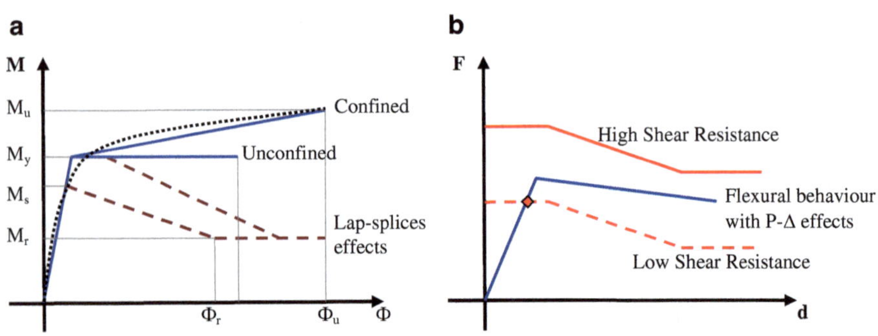

Fig. 4.42 Modelling of Piers: (**a**) Moment-curvature analysis of the critical section of the pier and (**b**) associated Force-displacement behaviour, taking into account possible shear failure

The mechanical behaviour of the bearing devices has been defined based on a survey of the Italian bridge stock. Five different types of bearing devices have been identified, i.e.: (i) Steel Hinges (SH), (ii) Sliding Bearings (S), (iii) Steel Pendulum and Roller Bearings (SP,SR), (iv) Neoprene Pads (NP), (v) Dowel Bars (DB). Steel hinges have been assumed to remain linear elastic up to failure. A post-failure frictional behaviour ($\mu_{fr} = 50\%$), corresponding to sliding between deck and pier cap, has been considered. Reference to a Coulomb (rigid-perfectly-plastic) model has been made to describe the frictional behaviour of S, SP and SR (see Fig. 4.41b), considering different friction coefficients (μ_{fr}) for each bearing type (Cardone et al. 2011). A linear visco-elastic behaviour has been assumed for neoprene pads (see Fig. 4.41b), with shear stiffness computed based on the dimensions of the pads and the shear modulus (G) of neoprene. A shear modulus of 1 MPa and a viscous damping ratio (ξ) of 6% have been assumed as typical values for neoprene pads. The horizontal strength of neoprene pads has been evaluated as the lowest value between the shear resistance of neoprene pads ($\gamma_{lim} = 150\%$) and the friction resistance between neoprene and concrete sliding surfaces ($\mu_{fr} = 70\%$).

4.3.4.4 Results

In Table 4.6 the Damage States (DS_{ij}), experienced by each bridge structure in the longitudinal and transverse direction, are described and the corresponding PGA values, derived with IACSM using the elastic spectrum of Eurocode 8 for soil type B (CEN 2004) as demand spectrum, are reported. As can be seen, in the longitudinal direction, the PGA values corresponding to the attainment of the first slight damage state (PL1) range between approximately 0.05 and 0.25 g and those corresponding to the first severe damage state (PL3) between approximately 0.2 and 1 g. In the transverse direction, instead, the PGA values corresponding to the attainment of the first slight damage state (PL1) range between approximately 0.16 and 0.7 g and those corresponding to the first severe damage state (PL3) between approximately 0.4 and 1 g. The examination of the damage states in the longitudinal direction clearly points out that the behaviour of joints and abutments can considerably affect the seismic response of multi-span simply supported deck bridges. Joints and abutments must be correctly modelled, in order to properly estimate the seismic vulnerability of existing bridges, especially when the seismic response of the bridge is governed by the displacement capacity of the bearing devices.

In the last column of Table 4.6, the PGA values on stiff soil, at the bridge site, having 10% probability of exceedance in 50 years are reported. The ratio between PGA_{PLi} and $PGA_{10\%/50y}$ provides a first (deterministic) measure of the seismic vulnerability of the bridge. As can be noted, in most cases, the values of $PGA_{10\%/50y}$ result lower than PGA_{PL3}. The only exceptions are represented by Macchione, Fiumarella and Serra dei Lupi bridges in the longitudinal direction, where the values of $PGA_{10\%/50y}$ result very close to PGA_{PL3}, indicating that extensive structural damage is expected to occur in case of ground motions with 500 years mean return period. Actually, a number of seismic retrofit measures, consisting in the strengthening of the piers and in the replacement of the bearing devices, are being undertaken for the three bridges under consideration. It is worth to observe that for all the bridges examined the values of $PGA_{10\%/50y}$ result greater than PGA_{PL1}, these latter being associated to either pier yielding or joint closure or failure of neoprene pads.

Table 4.6 Performance levels, damage states, and corresponding PGA values in the longitudinal and transverse direction of the examined bridges

		Longitudinal		Transverse		
Bridge	PL	DS	PGA (g)	DS	PGA (g)	PGA $_{10\%/50y}$
Lauretta	PL1	DS11: Closure of abt-deck joint	0.18	DS12: Pier yielding	0.25	0.34
		DS12: Pier yielding	0.28			
	PL2	DS21: 50% ultimate ductility of pier	0.50	DS21: 50% ultimate ductility of pier	0.38	
	PL3	DS34: Passive resistance of abutment	0.60	DS32: Pier ultimate ductility	0.50	
Ceraso	PL1	DS11: Closure of abt-deck joint	0.25	DS12: Pier yielding	0.23	0.34
		DS12: Pier yielding	0.35			
	PL2	DS21: 50% ultimate ductility of pier	0.45	DS21: 50% ultimate ductility of pier	0.34	
	PL3	DS32: Pier ultimate ductility	0.54	DS32: Pier ultimate ductility	0.42	
San Gennaro	PL1	DS11: Closure of abt-deck joint	0.093	DS11: Closure of deck-deck joint	0.18	0.25
		DS13: Failure of neoprene pads	0.30			
	PL2			DS22: Displ. capacity of sliders	0.48	
	PL3	DS34: Passive resistance of abutment	0.32	DS35: Deck unseating	0.70	
Castello	PL1	DS11: Closure of abt-deck joint	0.073	DS11: Closure of deck-deck joint	0.18	0.25
				DS13: Failure of neoprene pads	0.40	
	PL2			DS22: Displ. capacity of sliders	0.49	
	PL3	DS34: Passive resistance of abutment	0.37	DS35: Deck unseating	0.82	
Carapelle	PL1	DS11: Closure of deck-deck joint	0.078	DS11: Closure of deck-deck joint	0.23	0.25
		DS13: Failure of neoprene pads	0.25	DS13: Failure of neoprene pads	0.39	
	PL2			DS22: Displ. capacity of sliders	0.40	
	PL3	DS34: Passive resistance of abutment	0.48	DS35: Deck unseating	0.68	
Fiumarella	PL1	DS12: Pier yielding	0.15	DS12: Pier yielding	0.22	0.33
		DS11: Closure of deck-deck joint	0.16			
	PL2	DS21: 50% ultimate ductility of pier	0.25	DS21: 50% ultimate ductility of pier	0.45	
		DS23: Active resistance of abutment	0.26			
		DS22: Displ. capacity of steel pendulum	0.28			
	PL3	DS32: Pier ultimate ductility	0.33	DS32: Pier ultimate ductility	0.60	

Serra Lupi	PL1	DS11: Closure of abt-deck joint	0.19			0.33
	PL3	DS33: Pier shear failure	0.30	DS33: Pier shear failure	0.37	
Vallone Volpi	PL1	DS12: Pier yielding	0.13			0.26
		DS11: Closure of deck-deck joint	0.18			
	PL2	DS23: Active resistance of abutment	0.36			
	PL3	DS31: Failure of dowel bars	0.58	DS31: Failure of steel hinges	1.04	
Macchione	PL1	DS11: Closure of deck-deck joint	0.05	DS11: Closure of deck-deck joint	0.16	0.33
		DS12: Pier yielding	0.13	DS12: Pier yielding	0.30	
	PL2	DS22: Displ. capacity of steel roller	0.10	DS21: 50% ultimate ductility of pier	0.44	
		DS21: 50% ultimate ductility of pier	0.29			
	PL3	DS35: Deck unseating	0.31	DS32: Pier ultimate ductility	0.60	

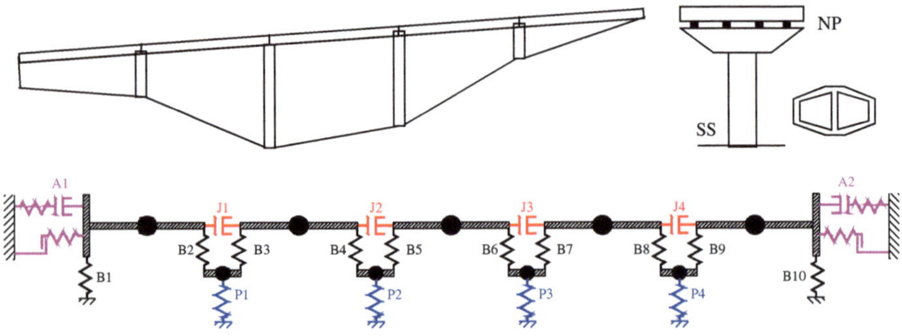

Fig. 4.43 Layout of the Ceraso viaduct and associated schematic model

Table 4.7 Tributary mass and force-displacement relationships of the piers of the Ceraso bridge

		Transverse				Longitudinal			
Piers	Tributary mass (ton)	F_y (kN)	d_y (mm)	F_u (kN)	d_u (mm)	F_y (kN)	d_y (mm)	F_u (kN)	d_u (mm)
P1	136.94	4,547.10	13.36	4,566.71	36.41	2,307.42	23.39	2,320.24	64.46
P2	155.38	2,597.27	40.95	2,608.06	109.02	1,317.98	71.68	1,325.30	192.85
P3	156.41	2,536.38	42.94	2,546.91	114.23	1,287.08	75.17	1,294.23	202.07
P4	140.32	3,996.82	17.29	4,013.89	46.83	2,028.18	30.27	2,039.44	82.872

Evaluation of Seismic Vulnerability

For the sake of brevity, herein the attention is focused on the Ceraso bridge only. The Ceraso viaduct is a 5-span simply-supported deck bridge with single shaft piers and neoprene bearings (see Table 4.4). Figure 4.43 shows the layout of the Ceraso bridge and its schematic model. Each deck has 33.5 m length and 720 ton mass. The joint width is 50 mm. Piers have a polygonal hollow cross section and effective heights ranging from 4.5 to 10 m.

In Table 4.7, the schematic bilinear force-displacement behaviour of each pier is outlined, separately in the transverse and longitudinal direction. The tributary mass of each pier is also reported.

The shear stiffness of the neoprene bearings has been derived based on their overall dimensions and the design value of the neoprene shear modulus. A horizontal stiffness of 20,000 kN/m has been thus assigned to each line of bearing devices. The horizontal strength of the bearing system has been computed as the lowest between the shear resistance of neoprene pads (1,604 kN) and the friction resistance between neoprene and concrete sliding surfaces (2,500 kN).

The longitudinal analysis of the bridge has been performed taking into account the effects of the pushing and pulling action of the deck on the abutments. In the pushing action, the abutment-backfill interaction determines a horizontal stiffness of 139,350 kN/m and a passive resistance of 6,000 kN. In the pulling action, the pile-soil interaction is modelled with a horizontal stiffness of 46,450 kN/m and an ultimate resistance of 5,000 kN.

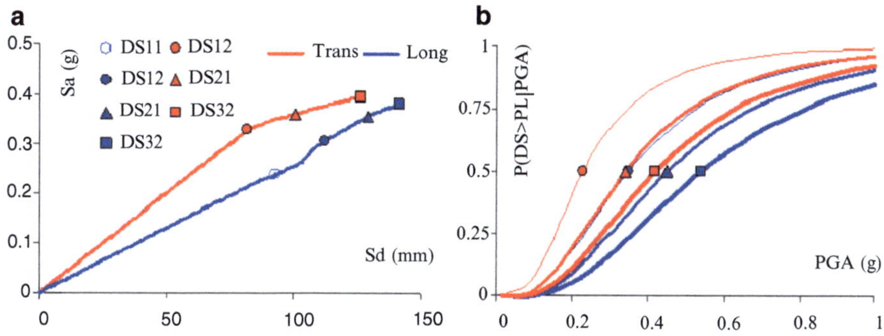

Fig. 4.44 (**a**) Capacity Spectra and Performance Points of the Ceraso bridge in the longitudinal and transverse directions; (**b**) Fragility Curves associated to the Damage States registered during *DAP* analysis

Table 4.8 Main results of the assessment procedure (Ceraso bridge, longitudinal direction)

DS	D_{CM} (mm)	S_d (mm)	V_b (kN)	S_a (g)	K_e (kN/m)	M_e (ton)	T_e (s)	ξ_e (%)	S_{a1} (T_e,ξ_e) (g)	PGA (g)
DS11	139.69	93.00	8,148.00	0.24	87,609.52	3,488.84	1.25	5.00	0.93	0.25
DS12	160.35	112.29	10,716.21	0.31	95,429.95	3,580.04	1.22	5.00	0.87	0.35
DS21	171.56	128.74	12,932.24	0.35	100,448.70	3,715.10	1.21	9.89	0.79	0.45
DS32	178.85	141.04	14,120.63	0.38	100,118.80	3,787.96	1.22	12.65	0.71	0.55

Table 4.9 Main results of the assessment procedure (Ceraso bridge, transverse direction)

DS	D_{CM} (mm)	S_d (mm)	V_b (kN)	S_a (g)	K_e (kN/m)	M_e (ton)	T_e (s)	ξ_e (%)	S_{a1} (T_e,ξ_e) (g)	PGA (g)
DS12	105.16	71.08	9,710.48	0.28	136,604.13	3,492.02	1.00	5.00	1.17	0.23
DS21	144.69	100.60	12,087.17	0.36	120,154.16	3,429.07	1.06	6.69	1.06	0.34
DS32	178.30	126.11	12,912.77	0.40	120,390.48	3,331.63	1.13	7.67	0.94	0.42

In Fig. 4.44a the capacity curves of the equivalent SDOF model of the Ceraso bridge are shown, separately in the longitudinal and transverse direction. On each curve, the PPs corresponding to the DSs that take place during DAP analysis (i.e. pier yielding, pier 50% ultimate ductility, and pier collapse) are identified. The associated PGA values are 0.35, 0.45, 0.55 g in the longitudinal direction and 0.23, 0.34, 0.42 g in the transverse direction. In Fig. 4.44b the corresponding fragility curves, expressed by a lognormal cumulative probability function, are reported. Tables 4.8 and 4.9 summarize the main results relevant to each DS in the transverse and longitudinal direction, respectively, i.e.: (i) the displacement of the centre of mass (D_{CM}), (ii) the displacement of the equivalent SDOF system (S_d), (iii) the total base shear (V_b), (iv) the acceleration of the equivalent SDOF system (S_a), (v) the equivalent stiffness (K_e), (vi) the effective mass (M_e), (vii) the equivalent period (T_e), (viii) the equivalent damping (ξ_e), (ix) the normalized spectral acceleration (S_{a1}) and (x) the Peak Ground Acceleration (PGA) associated to each DS.

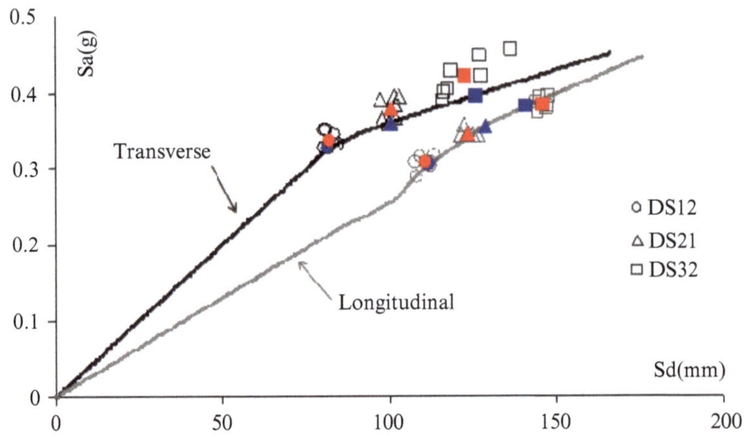

Fig. 4.45 Comparison between *NRHA* results and *IACSM* predictions in terms of maximum seismic response of the equivalent *SDOF* model of the Ceraso bridge in the longitudinal and transverse direction

Comparison with NRHA Results

Comprehensive nonlinear response-history analyses have been carried out to evaluate the accuracy of the proposed method (IACSM). The bridge displacement profiles based on IACSM have been compared with the envelope of the maximum bridge displacements (averaged on 7 accelerograms) obtained from NRHA. For the sake of brevity, herein the attention is focused on the Ceraso bridge only. Comprehensive results can be found in Cardone et al. (2011).

In Fig. 4.45 the NRHA results are expressed in terms of maximum acceleration vs. maximum displacement of the equivalent SDOF system and compared with the predictions of the IACSM. As can be seen, a good agreement between the average values of the NRHA results and the expected PPs is observed.

In Fig. 4.46, the seismic response of the Ceraso bridge in the longitudinal direction is examined. The average values of absolute deck displacements and relative joint displacements obtained from NRHA are compared with those predicted by the IACSM for three different DSs, corresponding to pier yielding, attainment of 50% pier ultimate ductility, and pier collapse, respectively. In Table 4.10 the comparison is made in terms of maximum pier top displacement and deck displacements.

As can be seen, in Fig. 4.46 two series of results are reported for IACSM. They differ in the verse of application of the lateral displacements in the DAP analysis: from left to right (IACSM$^{\rightarrow}$) and from right to left (IACSM$^{\leftarrow}$), respectively. The execution of two pushover analyses is fundamental for bridges with unsymmetrical configuration of piers and/or pier-deck connections. In these situations, reference to the envelope between IACSM$^{\rightarrow}$ and IACSM$^{\leftarrow}$ should be made. For the sake of brevity, in the following the acronym IACSM is used to identify the envelope of IACSM$^{\rightarrow}$ and IACSM$^{\leftarrow}$.

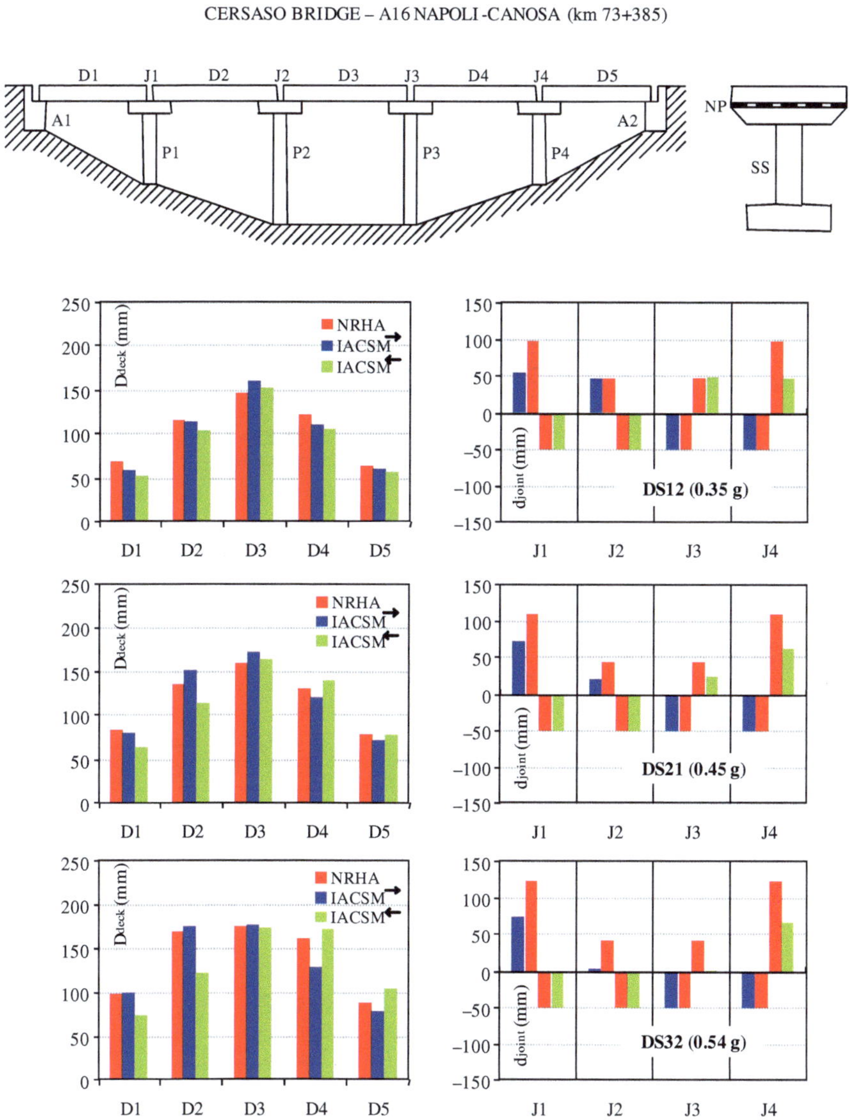

Fig. 4.46 Seismic response of the Ceraso bridge in the longitudinal direction. Comparison between IACSM predictions and NRHA results (maximum values averaged over 7 ground motions): (*left*) absolute deck displacements and (*right*) relative joint displacements

As far as joint displacements are concerned, it should be noted that negative values correspond to two adjacent decks that move closer together. The limit value in this case is represented by the joint width (50 mm), where pounding between decks occurs.

Table 4.10 Comparison between *IACSM* and *NRHA* – maximum displacements at the top of piers and at the decks in the longitudinal direction

	DS12			DS21			DS32		
Longitudinal direction	NRHA (mm)	IACSM (mm)	ERR (%)	NRHA (mm)	IACSM (mm)	ERR (%)	NRHA (mm)	IACSM (mm)	ERR (%)
D1	68.01	58.40	14.13	83.11	79.18	4.73	99.30	100.07	−0.78
P1	28.51	23.39	17.98	46.37	43.85	5.44	73.64	64.48	12.44
D2	114.81	113.26	1.35	135.94	151.06	−11.12	168.54	175.66	−4.22
P2	110.84	103.90	6.26	130.06	128.38	1.29	153.50	144.31	5.99
D3	147.00	160.35	−9.08	158.99	171.56	−7.91	176.31	178.85	−1.44
P3	106.40	103.20	3.00	120.73	120.25	0.40	142.94	140.70	1.57
D4	120.29	110.31	8.29	131.00	140.51	−7.26	161.00	172.31	−7.02
P4	33.43	34.56	−3.38	44.93	56.68	−26.15	62.57	83.04	−32.71
D5	62.99	60.02	4.71	77.47	78.05	−0.75	87.24	104.64	−19.94

Generally speaking, a good agreement between IACSM predictions and NRHA results is observed for the Ceraso bridge, for all the DSs considered. This is confirmed by Table 4.10, where the percent errors, calculated as the absolute difference between NRHA and IACSM results divided by the "exact" NRHA value, are reported. As can be seen, errors less than 20% are found in terms of deck displacements, with the IACSM tending to be a little conservative. Moreover, the level of accuracy of the IACSM is practically the same passing from slight damage states (DS12 in Fig. 4.46), where all the piers are still elastic, to severe damage states (DS32 in Fig. 4.46), where extensive plastic deformations occur. The good agreement between the two displacement profiles related to DS32 confirms that the procedure is able to capture the actual distribution of plastic deformations in the structure. In particular, both IACSM and NRHA identify pier no. 1 as the critical member of the bridge, where yielding, 50% ultimate ductility and collapse first take place. Also piers no. 2, 3 and 4 however, undergo extensive plastic deformations with ductility demands that differ from those expected based on the IACSM less than 30% (see Table 4.10), regardless of the DS considered.

As far as joint displacements are concerned, significant differences between IACSM predictions and NRHA results are observed. They can be ascribed to asynchronous movements between adjacent decks due to the higher-mode response of the bridge, that appears to be a little underestimated in the DAP analysis.

In Fig. 4.47, IACSM predictions and NRHA results are compared for the transverse direction of the Ceraso bridge. The comparison is made in terms of deformed shapes of the decks associated to three different DSs, corresponding to pier yielding (PGA = 0.23 g), attainment of 50% pier ultimate ductility (PGA = 0.34 g) and pier collapse (PGA = 0.42 g), respectively. In Table 4.11 the comparison is made in terms of maximum displacements of top piers and deck ends.

Based on the IACSM, the critical pier, where yielding, 50% ultimate ductility and collapse first take place, is the pier no. 3. Also pier no. 2 undergoes extensive plastic deformations while piers no. 1 and pier no. 4 remain elastic.

For the sake of clarity, in Fig. 4.47 transverse displacements and bridge coor dinates are reported in two different scales. From a graphical point of view, this determines a distortion of the deformed shape of the bridge that considerably amplifies the rotations of the decks. Nevertheless, the comparison between expected and 'actual' deformed shapes clearly points out the great accuracy of the IACSM in the prediction of the PGA values associated to slight-to-severe damage states. This is confirmed by Table 4.11, where the percent errors of the IACSM in the evaluation of the 'actual' maximum deck displacements are reported. As can be seen, they do not exceed 15% for DS12 and 25% for DS32 and, on average, they are around 9% for DS12 and around 13% for DS32. The NRHA results also confirm that pier no. 3 is the critical element of the bridge. Also pier no. 2 undergoes significant plastic deformations with ductility demands that differ from those expected based on the IACSM less than 15% (see Table 4.11), regardless of the DS considered.

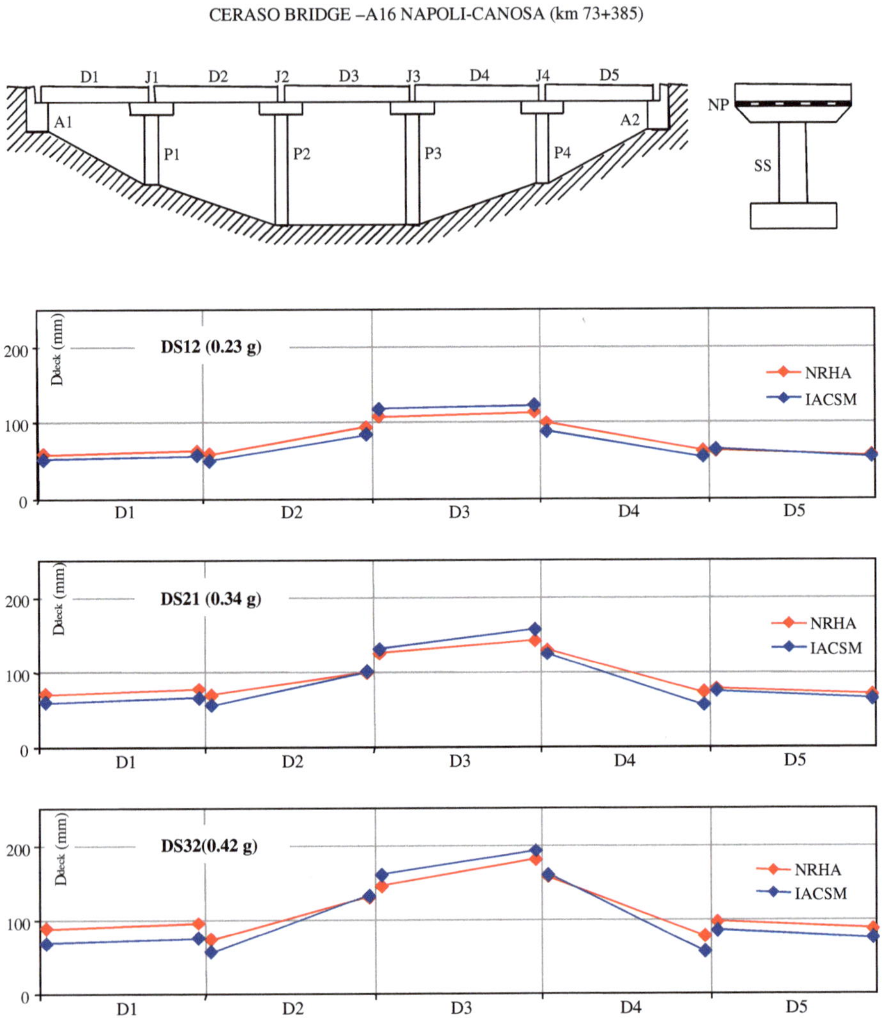

Fig. 4.47 Seismic response of the Ceraso bridge in the transverse direction. Comparison between *IACSM* predictions and *NRHA* results (maximum values averaged over 7 ground motions) in terms of deformed shape of the bridge

In order to better quantify the accuracy of the IACSM in capturing the 'exact' maximum deformed shape of the bridge, the following Bridge Index (BI) has been computed for each selected DS:

$$\mathrm{BI} = \mathrm{median}_{j=1\ldots 2Nd}\left(\frac{D_{j,IACSM}}{D_{j,NRHA}}\right) \tag{4.13}$$

Table 4.11 Comparison between *IACSM* and *NRHA* – maximum displacements at the top of piers and at the deck ends in the transverse direction

	DS12			DS21			DS32		
Transverse direction	NRHA (mm)	IACSM (mm)	ERR (%)	NRHA (mm)	IACSM (mm)	ERR (%)	NRHA (mm)	IACSM (mm)	ERR (%)
$D1_L$	58.36	51.93	−11.02	71.67	59.52	−16.96	88.89	68.88	−22.51
$D1_R$	63.21	57.61	−8.86	77.61	65.96	−15.02	96.37	75.89	−21.26
P1	6.03	5.68	−5.74	7.29	6.44	−11.65	8.43	7.01	−16.86
$D2_L$	58.87	50.46	−14.29	69.24	56.46	−18.46	74.21	57.37	−22.69
$D2_R$	94.27	84.08	−10.81	99.55	101.09	1.55	132.16	133.08	0.70
P2	41.04	39.30	−4.24	58.80	51.07	−13.15	79.06	82.71	4.63
$D3_L$	108.09	119.17	10.25	127.37	130.98	2.84	146.86	162.55	10.68
$D3_R$	114.34	122.70	7.31	143.54	158.40	10.35	182.13	194.06	6.55
P3	46.71	42.84	−8.29	80.09	78.48	−2.01	118.80	114.23	−3.85
$D4_L$	101.13	89.50	−11.50	129.49	125.64	−2.97	158.04	161.74	2.34
$D4_R$	63.73	55.54	−12.86	72.54	56.92	−21.54	77.34	58.22	−24.72
P4	8.99	8.88	−1.20	10.77	9.75	−9.46	12.26	10.71	−12.65
$D5_L$	63.70	64.83	1.77	78.26	75.30	−3.78	97.63	86.93	−10.96
$D5_R$	57.06	55.95	−1.94	70.13	65.55	−6.53	87.63	76.23	−13.01

Table 4.12 Bridge Index (*BI*) of the bridges examined for different damage states

	Longitudinal		Transverse	
Bridge	DS	BI	DS	BI
Lauretta (SS/NP)	DS12	1.03	DS12	1.06
	DS21	0.92	DS21	0.98
	DS34	0.91	DS32	0.91
Ceraso (SS/NP)	DS12	0.95	DS12	0.9
	DS21	1.07	DS21	0.95
	DS32	1.04	DS32	0.88
San Gennaro (SW/NP + SB)	DS11	0.87	DS11	1.01
	–	–	DS22	1.06
	DS34	1.02	–	–
Castello (SW/NP + SB)	DS11	0.98	DS11	1.02
	–	–	DS22	1.45
	DS34	1.11	–	–
Carapelle (SW/NP + SB)	DS11	0.95	DS11	1.03
	–	–	DS22	0.86
	DS34	1.06	–	–
Fiumarella (SF/SH + SP, SH)	DS12	0.94	DS12	1
	DS21	1.10	DS21	0.87
	DS32	1.10	DS32	0.61
Serra Dei Lupi (SF/DB + SR, DB + SH)	DS33	1.10	DS33	1.01
Vallone Delle Volpi (SW/DB + SR, DB + SH)	DS12	1.18	–	–
	DS23	0.94	–	–
	DS31	1.14	DS31	0.90
Macchione (SF/SH + SR, SH)	–	–	DS12	0.99
	–	–	DS21	0.98
	DS35	1.06	DS32	0.93

where $D_{j,IACSM}$ is the displacement of the j-th deck end provided by IACSM, $D_{j,NRHA}$ is the corresponding maximum displacement (averaged over 7 accelerograms) derived from NRHA and N_d is the number of decks. The ideal target value of BI is always 1.

In Table 4.12 the values of BI, for the selected DSs are reported, separately for the transverse and longitudinal direction of the nine bridges examined. In the longitudinal direction, the BI ranges between 0.87 and 1.18. In the transverse direction (excluding a few cases), the BI ranges between 0.89 and 1.05. In the 80% of the totality of cases examined the BI results between 0.9 and 1.1.

The lowest value of BI in the transverse direction is related to the DS32 of the Fiumarella bridge, for which the IACSM significantly underestimates the 'actual' maximum deformed shape of the bridge, since it does not predict correctly the inelastic behaviour of piers. The highest value of BI in the transverse direction is registered for the DS22 of the Castello bridge. In this case the IACSM significantly overestimates the 'actual' maximum deformed shape of the bridge, since it is essentially governed by the friction behaviour of the sliding bearings, with equivalent damping ratio greater than 40%, which is beyond the limit of applicability of Eq. 4.11.

4.3.4.5 Concluding Remarks

A performance-based methodology for the seismic assessment of highway bridges has been presented. The proposed methodology is based on an Inverse Adaptive application of the Capacity Spectrum Method, referred to as IACSM. The IACSM provides the PGA values associated to predefined Damage States (DSs) of the critical members of the bridge (piers, abutments, bearing devices, joints). Based on these PGA values, a number of fragility curves are derived to describe the seismic vulnerability of the bridge from a probabilistic point of view.

Here, the IACSM has been applied to a set of nine simply-supported deck bridges of the Italian A16 highway. The predictions of the IACSM have then been evaluated against the results of Nonlinear Response-History Analyses, carried out using a set of seven accelerograms, compatible with the EC8-Ground B response spectrum, scaled to the PGA values provided by IACSM for the selected DSs. The comparison has been made in terms of maximum deformed shapes of the deck, joint displacements, and top pier displacements.

The NRHA results clearly confirm the good accuracy of the proposed methodology in predicting the PGA values associated to slight-to-severe DSs, for several bridges structures, differing in pier types and pier-deck connections.

4.4 Experimental Evaluation of Analytical Methods

In the previous sections the various types of pushover methods were evaluated against the results of NRHA, assumed to be the ‘exact’ solution. In this last section, various inelastic analysis methods are evaluated against actually measured response during testing bridge structures. It is recalled here that the number of studies of the applicability of the pushover methods for the analysis of bridges is still small if compared to those performed on buildings. Previous investigations (Isaković and Fischinger 2006; Isaković et al. 2008; Paraskeva et al. 2006; (Lupoi et al. 2007; Pinho et al. 2007) were mostly related to the single-column bent viaducts and were mostly analytical.

4.4.1 Applicability of Analytical Methods to the Seismic Analysis of an RC Bridge, Experimentally Tested on Three Shake Tables

4.4.1.1 Introduction

Within the research, presented in this section, the findings of the analytical studies were tested by means of the experimentally observed response of a typical RC

Fig. 4.48 The pre-NEES experiment performed at the University of Nevada, Reno (courtesy of the *UNR*)

bridge. Three typical methods investigated in the previous studies were employed: (a) the N2 method (Fajfar and Fischinger 1987; Fajfar et al. 1997) as a typical single mode pushover method, which is included into the EC8, (b) the MPA method (Chopra and Goel 2002), as a typical non-adaptive multimode pushover method, and (c) the IRSA method (Aydinoglu 2004) as a typical adaptive multimode method.

The experiment, which was used to examine the applicability of these pushover methods, was performed in the frame of the NEES project "Seismic Performance of Bridge Systems with Conventional and Innovative Materials" (UNR 2008), coordinated by Prof. M. Saiidi from the University of Nevada, Reno (UNR). Several tests on three parallel shake tables have been performed in the frame of this project. One of the main purposes of the project has been to verify the computer simulations by means of experimental data in order to establish the reliability of the analytical studies. Only the so-called pre-NEES experiment is studied in this section (see Fig. 4.48). A two-span two-column bent bridge, typical of US practice was analyzed.

This section includes:

(a) a brief description of the main properties of the analyzed bridge,
(b) description of the related analytical model, which was used in the analytical studies,
(c) the summary of the experimentally observed bridge response,
(d) the comparison of experimental and analytical results.

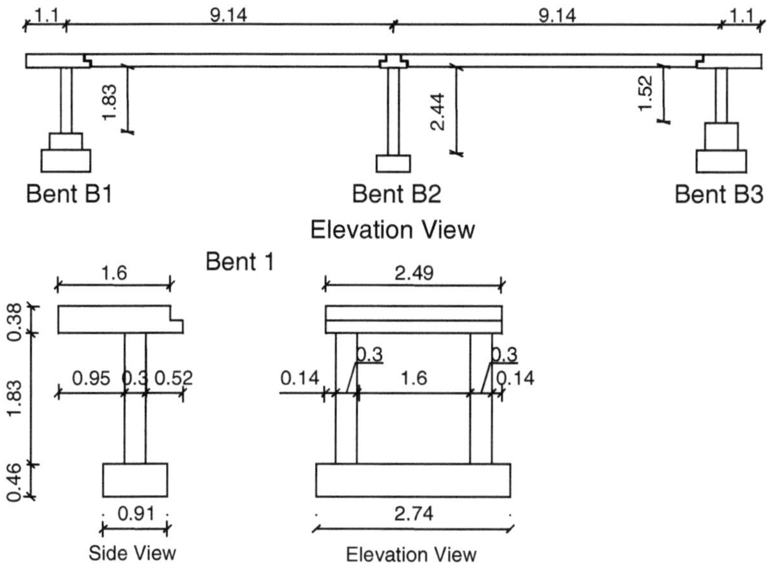

Fig. 4.49 The scheme of the investigated bridge

4.4.1.2 Short Summary of the Experiment

The analyzed bridge is a moderately irregular structure, representing one frame of the continuous bridge, typical of the US practice. The 1:4 scale model is presented in Fig. 4.49. This is a two-span bridge, supported by three two-column bents. The length of the individual span was 9.14 m. The bridge model was supported by six columns of diameter 30.5 cm. The height of columns was 1.83, 2.44 and 1.52 m for the bents B1, B2 and B3, respectively. The geometry of the bent B1 is presented in Fig. 4.49. Columns were reinforced by 16 bars of diameter $\phi 9.5$. The transverse reinforcement consisted of the spiral reinforcement of diameter $\phi 4.9$. The space between transverse bars was 32 mm.

The bridge was subjected to different seismic intensity levels. The main investigation was related to seven high-amplitude tests, noted as tests T12–T19. These tests (except the last one) were included in the analytical studies. The intensity of the table accelerations were varied between 0.08 (T12) and 2.11 g (T19). Due to the interaction between the structure and the shake tables, the excitations at the tables were somewhat different. Typical acceleration and displacement spectra corresponding to the accelerations applied during the test T14 are presented in Fig. 4.50. More detailed description of the bridge and the applied load can be found in Johnson et al. (2006, 2008).

4.4.1.3 Analytical Model

An analytical model of the bridge is schematically presented in Fig. 4.51. Superstructure was modelled using elastic beam-column elements. The nonlinear flexural

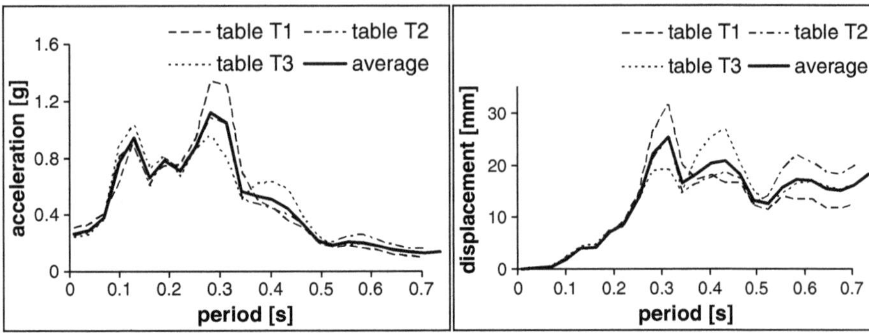

Fig. 4.50 The typical acceleration and displacement spectra (test T14; average PGA = 0.31 g)

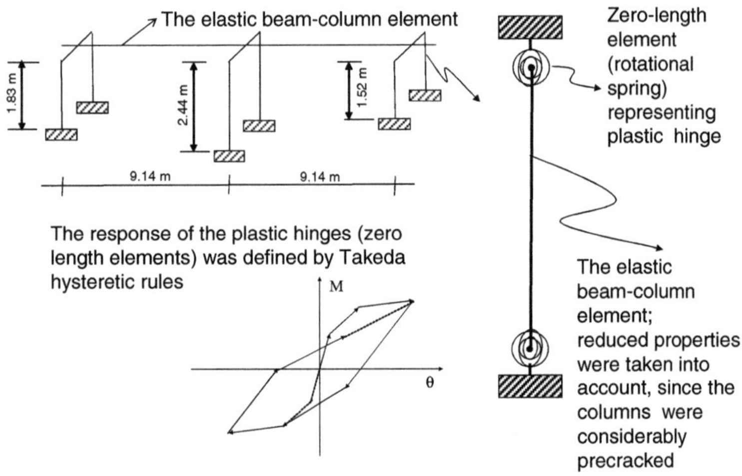

Fig. 4.51 Analytical model of the bridge

response of columns was modelled using zero length elements at both ends of columns, which were linked by elastic beam-column elements. The nonlinear behaviour was defined by Takeda's hysteretic rules, as implemented in the OpenSees program system (Mazzoni et al. 2006) by Japanese researchers (Takahashi 2009). The detailed description of the analytical model could be found elsewhere (Isaković and Fischinger 2011).

4.4.1.4 Main Properties of the Experimentally Observed Response That Influenced the Application of the Pushover Methods

The main characteristics of the seismic response of the investigated bridge, which influenced the applicability of the pushover methods, were: considerable variations

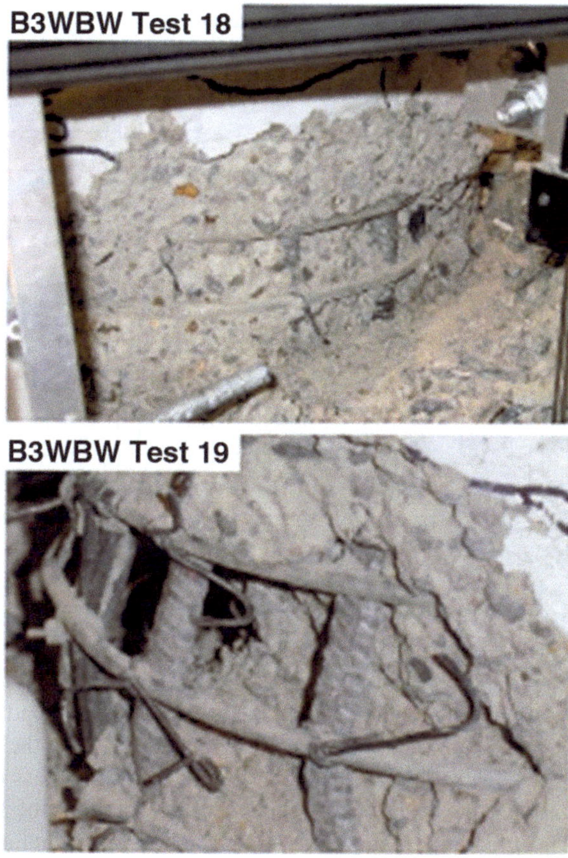

Fig. 4.52 Damage of one of the columns of bent B3 during the tests T18 and T19 (courtesy of the *UNR*)

of the deck rotations when the seismic intensity was changed, significant damage of the side bents, slightly different excitations of different shake tables.

The response of the analyzed bridge was predominantly influenced by the ratio of the stiffness of the side bents (B1 and B3). This ratio was changing depending on the different progress of the damage of bents B1 and B3. During the tests with lower seismic intensities (T12–T14) bent B1 was more exposed and its effective moment of inertia was smaller than that of bent B3. However, during the medium seismic intensity test (T15) bent B3 was damaged more than bent B1 and consequently the effective moment of inertia of bent B3 was now smaller than that of bent B1. Significant damage and failure of some longitudinal bars was observed in columns of bent B3 during the test T18, and final failure of bent B3 was reached during test T19 (see Fig. 4.52).

The ratio of effective moment of inertia (hence stiffness) of the side bents strongly influenced the mode shapes of the bridge. Therefore, a detailed study of these variations was performed. It is schematically presented in Fig. 4.53.

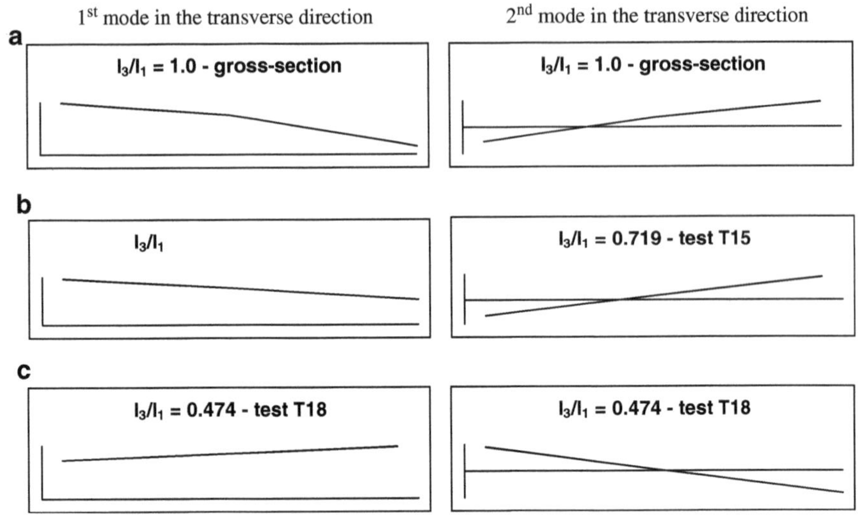

Fig. 4.53 Changes of the mode shapes depending on the ratio of the stiffness of the bents B1 and B3

When the columns of the bents B1 and B3 were uncracked, the moments of inertia of side columns were the same ($I_3/I_1 = 1$) and the ratio of the bent stiffness k_3/k_1 (k_1 and k_3 is the stiffness of the left (B1) and the right bent (B3), respectively) was 1.75 (see Fig. 4.53a). The centre of the stiffness of the bridge was at the right half of the bridge. Consequently, the rotations of the bridge were clock-wise. The first mode was predominant. The corresponding effective mass was about 84% of the total mass of the bridge.

When the seismic intensity was increased, the clock-wise rotations of the deck were firstly increased, and consequently the damage of bent B1 was proportionally increased. After the yielding of bent B1, bent B3 gradually became more exposed and the stiffness ratio of these side bents was changed. The stiffness of bent B3 was decreased compared to bent B1. Consequently the centre of stiffness was moved closer to the centre of the mass and the rotations of the deck were decreased (see Fig. 4.53b). The importance of the higher modes was significantly reduced. The response of the bridge was influenced mostly by one predominant mode.

At the test T18, bent B3 was more damaged than bent B1 and the centre of the stiffness was moved to the left part of the viaduct, changing the direction of the deck rotations (see Fig. 4.53c). The response was influenced by one predominant mode. During this test all bents were heavily damaged. For example, in the bent B3 buckling of some bars was observed (see Johnson et al. 2006). Due to considerably reduced column stiffness, the response of the bridge was quite sensitive to relatively small variations of column properties. The direction of the rotation was considerably changing depending on the ratio of the bents' stiffness. When bent B3 was damaged more than bent B1 the rotations of the deck changed the direction causing

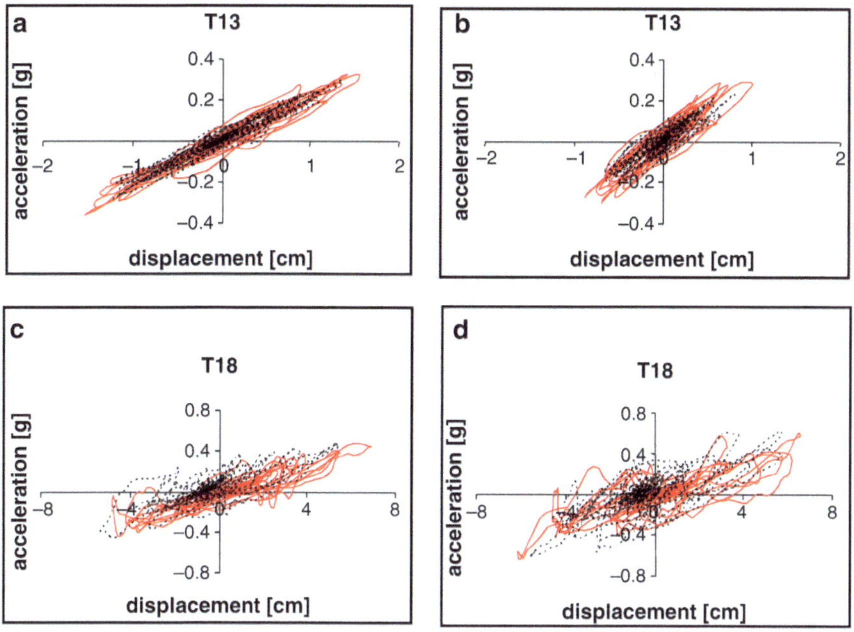

Fig. 4.54 (**a**) Experimental (*solid line*) and analytical (*doted line*) response of bent B1 during the test T13; (**b**) Experimental and analytical response of bent B3 during the test T13; (**c**) Experimental and analytical response of bent B1 during the test T18; (**d**) Experimental and analytical response of bent B3 during the test T18

larger displacements at the side of bent B3. Due to the large sensitivity of the response to small variations of the bent stiffness, the numerical modelling and the estimation of the response was quite a demanding task.

Different excitations of different shake tables complicated the application of the pushover methods, since only one spectra could be applied. Among different possibilities, that were considered in the analysis (for more details see Isaković and Fischinger 2007, 2011), it was found that when the average spectra was used, the best match of the results of the pushover and the nonlinear response history analyses was obtained.

4.4.1.5 Comparison of Analytical and Experimental Results

Results of the Nonlinear Response History Analysis (NRHA)

Using the NRHA the analytical model of the bridge was verified, first. Although the model was quite simple, the match between the analytical and experimental results was good for all seismic intensity levels. An example of the experimentally and analytically obtained hysteretic response is presented in Fig. 4.54.

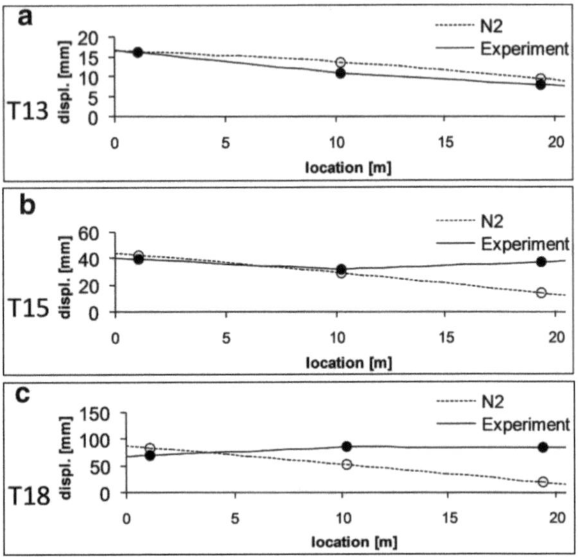

Fig. 4.55 Displacements calculated by the N2 method and displacements measured during the experiment

Results of the Single-mode Pushover Method

The results of the N2 method are compared with the experiment in Fig. 4.55. Only the tests, which illustrate the qualitative changes of the bridge response are presented (T13, T15 and T18); more details can be found elsewhere (Isaković and Fischinger 2011).

The accuracy of the displacement shapes depends on the seismic intensity. For the lower seismic intensities (tests T12–T14) the analytical displacement shapes correlated with the experimental data quite well (see test T13 in Fig. 4.55). When the load level was increased, the match of the displacement shapes was worse, particularly during and after test T15 when the qualitative changes of the bridge response were observed (see Fig. 4.55 and next sections). Due to the changed ratio of the side bents, the rotations of the bridge changed their direction. Several reasons for these qualitative changes were identified. The significant damage of the side bents during the tests T17 and T18 makes the response quite sensitive to small variations of the bent properties. Since the bents were heavily damaged, their resistance was significantly reduced in both the transverse and longitudinal direction. Consequently they were not capable to stabilize the rotations, unlike in buildings with undamaged elements perpendicular to the loading direction. The spectral characteristics of the applied seismic load caused that bent B3 was relatively more exposed than bent B1, as explained in the next sections.

When the bridge was analyzed by the N2 method, the direction of the rotations was always the same and independent of the load intensity. Contrary to the experimentally observed response, the rotations were always clock-wise and more pronounced under the strong excitations. The N2 method was not able to identify the

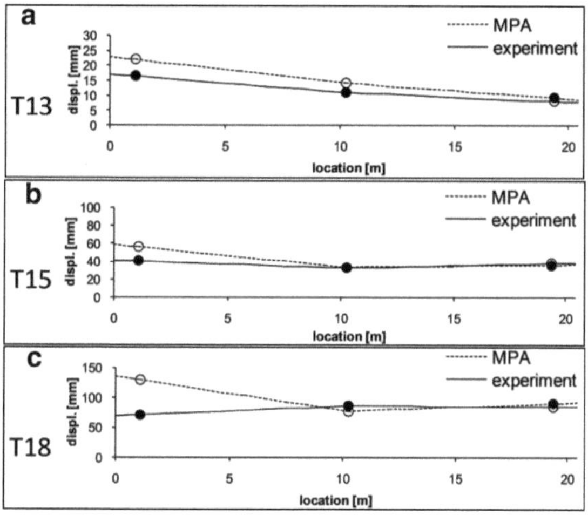

Fig. 4.56 Displacements calculated by the *MPA* method and displacements measured during the experiment

qualitative changes of the rotations of the superstructure. Consequently, the displacements of the right part of the viaduct (between bents B2 and B3) were considerably underestimated. The largest discrepancy was observed during test T18.

The results of the N2 method was somewhat improved when the extended version of the N2 method was applied, however the discrepancy was still considerable for higher seismic intensities (Isaković and Fischinger 2011).

Results of the Multi-mode Pushover Methods

The multimode methods were applied considering two modes in the transverse direction. At the initial investigated seismic level the effective mass was 84% and 16% for the first and the second mode, respectively. The contributions of these two modes were combined using the SRSS rule.

The displacements of the superstructure obtained with the MPA and the IRSA methods are compared with the experimental results in Figs. 4.56 and 4.57. The results for tests T13, T15 and T18 are presented, since they illustrate all basic characteristics of the response.

For the low seismic intensity levels correlation with the experiment was good for both multimode methods. The displacement shapes were estimated by the MPA method fairly well for the stronger earthquakes, too. Compared to the N2 method, the displacements shapes were considerably improved. However, detailed analysis of the response showed that the MPA has not identified the real cause of the change in the direction of the rotation.

During the experiment the response of the bridge was influenced predominantly by the first mode ($m_{eff} > 80\%$), which was considerably changing. The corresponding rotations changed the direction depending on the seismic level and during the test e.g. T15 and T18 the rotations of the deck were opposite to that in the tests with the lower seismic intensity. The non-adaptive MPA method could not take into account these

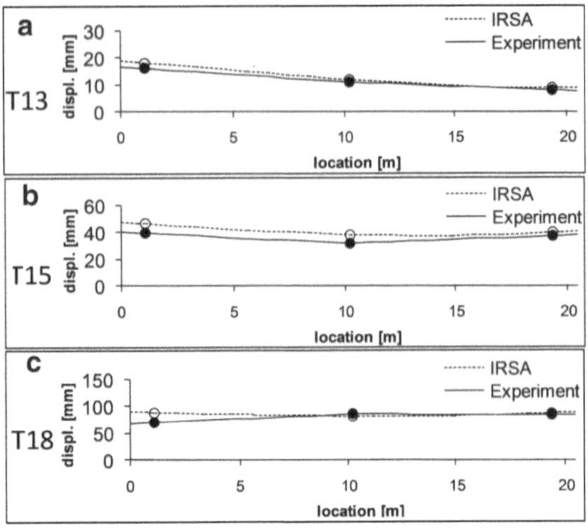

Fig. 4.57 Displacements calculated by the *IRSA* method and displacements measured during the experiment

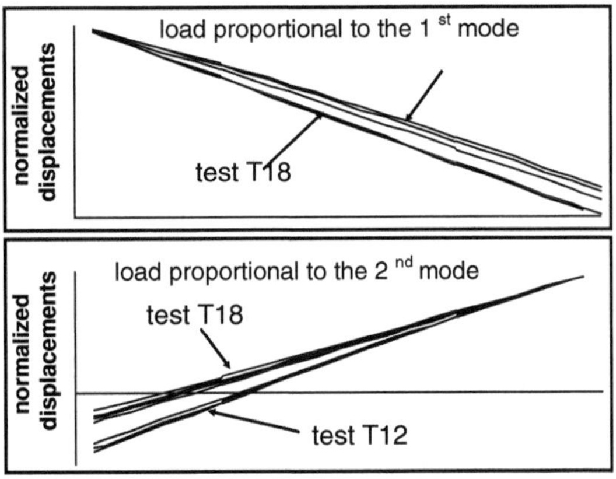

Fig. 4.58 Normalized displacements of the bridge calculated with the MPA

considerable changes of the predominant mode shape. Similar to the N2 method, the rotations of the bridge, subjected to inertial forces proportional to the first mode, were always clock-wise and they increased with the load intensity (see Fig. 4.58). However, the rotations corresponding to the inertial forces proportional to the second mode were in opposite direction (see Fig. 4.58). These rotations compensated for the difference between the experiment and the results of the analysis with the forces proportional to the first mode. Therefore, the total displacements match the experimental values reasonably well.

The adaptive IRSA method (Aydinoglu 2004) successfully identified changes of the important mode shapes, as well as the variable importance of the higher modes (see Fig. 4.57). The shape of the displacement line matched the experimental data quite well.

It should be noted that in all methods the absolute values of the displacements were quite sensitive to small variations in stiffness (of the analytical models), since the response spectra of the applied load were not smooth. When a slightly different model was used for the analysis, the absolute values of the displacements were different, but the displacement lines were qualitatively the same in both cases.

4.4.1.6 Concluding Remarks

A large scale model of two-span two-column bent RC bridge, representing one frame of typical US continuous bridge was experimentally tested on three parallel shake tables at the UNR in the frame of the NEES project. Although the geometry of the bridge was relatively regular, the experimentally observed response of the bridge was quite complex and considerably influenced by the seismic intensity and the ratio of the stiffness of the side bents. At the lower seismic intensities, the left (initially more flexible) side of the bridge was more exposed, resulting in the clockwise rotations and larger displacements at the left side of the bridge. At higher seismic intensity levels, the right (initially stiffer) bent was more exposed. Due to the damage of this bent the direction of the bridge rotations gradually changed, resulting in larger displacements at the right (initially stiffer) side. Several reasons for these qualitative changes of the response were detected: the significant damage of the side bents, which makes the response quite sensitive to small variations of the bent properties, the reduced resistance of columns in both (longitudinal and transverse) directions as well as the spectral characteristics of the applied seismic load.

Despite the fact that a rather simple numerical model of the bridge was used in the analytical studies, the results of nonlinear response history analysis matched the experiment quite well. Three typical pushover methods, which were capable to estimate both the capacity and the demand, the N2 method, the MPA method, and the IRSA method, correlated with the experiment quite well when the lower intensity levels were considered. The N2 method was less effective in the case of higher intensity levels since it was not able to take into account qualitative changes of the deck rotations.

The MPA estimated the displacement shapes quantitatively fairly well for the higher seismic intensities, too. However the detailed analysis of the response detected considerable qualitative differences with the experiment. The non-adaptive MPA method, could not take into account considerable changes of the first mode shape. The second mode coincidentally compensated this discrepancy and displacements of the right side of the bridge matched the experimental values well.

The IRSA method successfully identified the changes of the mode shapes as well as the variable importance of the higher modes. The analytical displacement shapes of the deck correlated with the experiment well.

4.4.2 Numerical Studies of RC Bridge, Supported by Hollow Box Columns Tested Pseudo-dynamically

4.4.2.1 Introduction

This section presents the seismic analysis of the Talübergang (ravine) Warth bridge studied within the framework of the European research project titled VAB – Vulnerability Assessment of Bridges (Faria et al. 2001). This bridge was built in Austria during the 1970s, designed to a very low seismic level, and consists of a

Fig. 4.59 General view of the pseudo-dynamic tests at the JRC-Ispra, 1:2.5 scale model (courtesy of the JRC-Ispra)

seven span continuous deck supported on two abutments and six rectangular hollow section piers, the latter with some peculiar characteristics concerning the reinforcement detailing whose modelling is quite demanding for cyclic response simulation. Still in the VAB project context, a physical scaled model of the bridge was also tested under pseudo-dynamic conditions at the JRC-Ispra (Pinto et al. 2002), illustrated in Fig. 4.59, and the results were compared with numerical simulations, as described in the following.

The non-linear behaviour is considered concentrated in the piers, which are discretized with (i) a refined constitutive model or (ii) a plastic hinge type model for the nonlinear material behaviour simulation. For the numerical prediction of the seismic performance of the Warth bridge these methodologies were adopted with the seismic action taken either as an asynchronous or synchronous ground motion induced along the transverse direction only.

The main results of the seismic analyses will be presented focusing on the essential role that the longitudinal reinforcement curtailment plays on macro-crack localization, which leads to a shift of the plastic hinge (usually at the base of piers) upwards, where a significant reduction of the longitudinal reinforcement takes place. From the comparison of the numerical predictions with the experimental results, as recorded during the pseudo-dynamic tests performed at the JRC, the capability of the damage model to provide accurate simulations of the seismic performance of the bridge was brought into evidence, despite the fact that the piers were difficult to simulate due to the concrete hollow section geometry and the unusual reinforcement layout adopted in the design.

The difficulties in carrying out analyses with methodologies that adopt hysteretic non-linear material behaviour increase significantly with the complexity of the model, involving a compromise between accuracy and computational cost, and with several parameters that have to be defined. Here a comparative study with several strategies to evaluate the seismic behaviour of bridges was envisaged. Therefore, two numerical models were used:

(i) PNL (Delgado et al. 2002) – nonlinear behaviour lumped in plastic hinges.
(ii) Damage Model (Faria et al. 1998) – refined constitutive model for the nonlinear material behaviour.

The numerical prediction of the seismic performance of this bridge adopts a 'peculiar' two-dimensional modelling of the entire bridge (Delgado et al. 2002; Vila Pouca et al. 2002) consisting of a simplified plane model intended for practical application and involving reduced computational effort, while maintaining adequate accuracy. The bridge structural modelling is carried out with plane elements, bars or 2D finite elements, maintaining the essential features of the 3D transverse response through an appropriate plane structural simulation. This is achieved by considering the deck and piers on the same plane, while the horizontal displacement of the pier top and the correspondent horizontal transverse deck displacement are constrained to have the same value by uniaxial tie elements. The comparative analyses were carried out for increasing levels of the seismic action, in order to show the sensitivity of the results to the different levels of non-linear response.

4.4.2.2 Description of the Bridge

The Talübergang Warth bridge studied by Faria et al. (2001) and Pinto et al. (2002) is illustrated in Fig. 4.60. This bridge consists of a deck with seven spans and a total length of 459 m, and is supported by six piers and two abutments; pier nomenclature is also indicated in Fig. 4.60.

Figure 4.61 and Table 4.13 report the basic geometry of the piers, both for the concrete section and for the footings (pier height L refers to the distance between the top of the footing and the base of the deck bearings); all piers have $6.8 \times 2.5\ m^2$ RC hollow sections.

In what concerns the curtailment of the longitudinal reinforcement, involving bars of different diameters, 4 regions are depicted in Fig. 4.61. Details of the reinforcement layout are provided in Faria et al. (2001) and Vila Pouca (2001). For piers P2, P3 and P4 the first interruption of the longitudinal rebars occurs close to the foundation, leading to a reduction of about 50% in the amount of steel reinforcement.

Transverse reinforcement is provided in the form of a single rectangular stirrup on each wall of the hollow section (Faria et al. 2001; Vila Pouca 2001), according to the following distribution: (i) $A_{sw} = \phi 12/20$ cm on the first 1 m layer close to the piers footing, and (ii) $A_{sw} = \phi 8/20$ cm elsewhere.

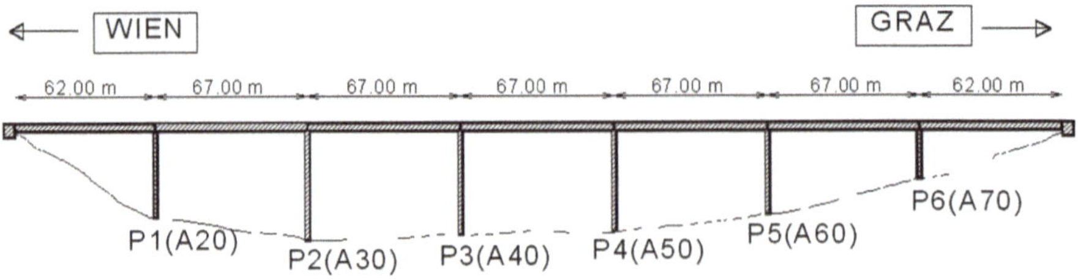

Fig. 4.60 Layout of Talübergang Warth bridge and pier nomenclature

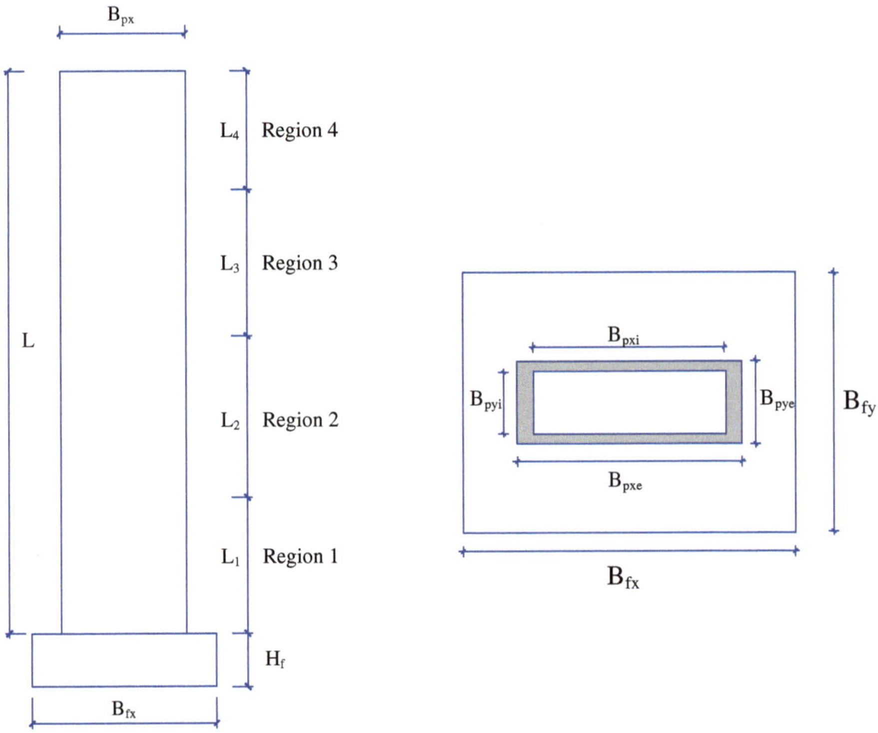

Fig. 4.61 Basic geometry nomenclature

Taking into consideration the stirrup arrangement documented in the design drawings, it is doubtful that concrete could benefit from any significant confinement. Therefore, for the Warth bridge piers, class B400 concrete (40 MPa design compressive strength) was assumed as unconfined, with the material properties reproduced in Table 4.14 (average values). Table 4.15 summarises the average material properties assumed for the longitudinal and transverse rebars, defined as corresponding to class RT50 steel.

Table 4.13 Bridge basic geometry

					Foundation		
Pier	L (m)	L/B_{pxe}	L/B_{pye}	Hollow section	B_{fx}	B_{fy}	H_f
P1 (A20)	29.8	4.4	11.9		10.8	10.1	3.45
P2 (A30)	38.9	5.7	15.6	$B_{pxe} = 6.8$ m	10.2	8.0	2.80
P3 (A40)	37.8	5.6	15.1	$B_{pxi} = 5.8$ m	10.2	8.0	2.80
P4 (A50)	36.0	5.3	14.4	$B_{pye} = 2.5$ m	10.2	8.0	2.80
P5 (A60)	30.0	4.4	12.0	$B_{pyi} = 1.9$ m	10.5	9.0	3.20
P6 (A70)	16.9	2.5	6.8		10.4	9.5	3.20

Table 4.14 Concrete properties (class B400, unconfined)

E (GPa)	f_{co} (MPa)	ε_{co}	f_o^+ (MPa)	f_{cm} (MPa)	ε_{cm}	Z
33.5	43.0	2.0‰	3.1	–	–	100

Table 4.15 Steel properties (class RT50)

f_{sy} (MPa)	f_{su} (MPa)	ε_{su}	ε_{sh}	E_h/E	R_o	a_1	a_2
545	611	100‰	5.0‰	0.0034	20	18.5	0.15

f_{su} and ε_{su} are the ultimate strength and strain for the steel

4.4.2.3 Simulation of the PSD Test (Asynchronous Motion)

The Warth bridge was tested in the ELSA Laboratory (Joint Research Centre, at Ispra) under pseudo-dynamic conditions and asynchronous ground motion induced along the transverse direction only; all experimental results are presented in the reference Pinto et al. (2002).

The non-linear behaviour is considered concentrated in the piers, which are discretized with a refined constitutive model for the nonlinear material behaviour. The refined numerical model combines a 2D plane-stress finite element discretization for concrete with 2-noded truss elements to model steel reinforcement. Concrete behaviour is modelled using a constitutive law based on Continuum Damage Mechanics, involving two independent scalar damage variables to account for degradation under tensile and compressive stress conditions. The cyclic response of steel is simulated via the Giuffré-Menegotto-Pinto model (Giuffrè and Pinto 1970).

The numerical analysis was performed in a sequential way, where the accelerograms were input with increasing intensities, in order to better reproduce the experimental tests; when a given seismic action is considered the structure has already been modified due to the non-linear effects induced by the previous one.

Figure 4.62 illustrates the comparison of the pier P3 top displacement responses obtained numerically (using the Damage Model) and experimentally (pseudo-dynamic test performed at the JRC), for three different levels of earthquake intensity. These three responses are in general close to the measured ones with respect to both frequency and maximum amplitude, except for the high-level earthquake where some differences are detected in last part of the response.

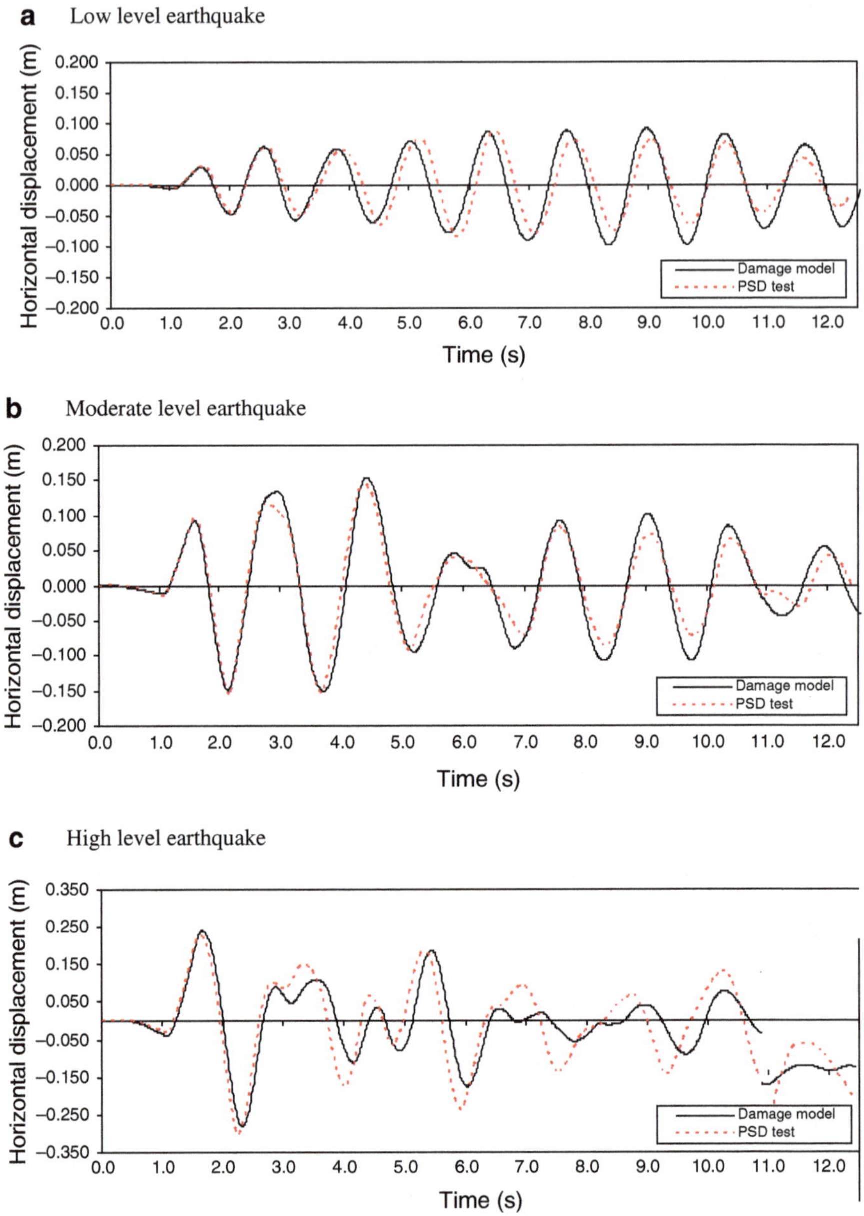

Fig. 4.62 Top horizontal displacement history of pier P3, comparison between *PSD* test and damage model. (**a**) Low level earthquake; (**b**) Moderate level earthquake; (**c**) High level earthquake

From the comparison of the numerical simulations with the experimental results, the capability of the damage model to provide accurate simulations of the seismic performance of the bridge was brought into evidence, despite the fact that the piers are difficult to model due to the concrete hollow section geometry and to the unusual reinforcement layout adopted in the design. In fact, the longitudinal reinforcement curtailment represents an important influence on macro-crack localization, which leads to a shift of the plastic hinge upwards where a significant reduction of the longitudinal reinforcement takes place.

4.4.2.4 Synchronous Numerical Analysis

In this subsection, numerical analyses were carried out assuming synchronous ground motion in the transverse direction, considering the pier P3 accelerogram (of the asynchronous case) as the input in this case.

Again, numerical analysis was performed in a sequential way, where the accelerograms were input with increasing intensities, in such a way that when a given seismic action is considered the structure has already been modified due to the non-linear effects induced by the previous one. The non-linear behaviour was considered concentrated in the piers, which are discretized with (i) the Damage model, a refined constitutive model, or (ii) the PNL model (Delgado et al. 2002; Delgado et al. 2004), a plastic hinge type model for the nonlinear material behaviour.

Since the capability of the damage model to provide accurate simulations of the bridge seismic performance was verified from previous asynchronous responses, it is reasonable to assume that accurate response simulations are expected for synchronous input motion, too.

Figure 4.63 illustrates the comparison of displacement responses of the pier P3 top, obtained numerically, using the Damage Model and the plastic hinge model, for three different intensity level earthquakes.

Assuming as the synchronous ground motion of the bridge the acceleration response of pier P3 (used in asynchronous analysis), the comparison between the asynchronous and synchronous response obtained with the damage model, Figs. 4.62 and 4.63, allows to verify that maximum peak displacement is reasonably similar in the two models, for moderate and high level earthquake. Both responses are similar at the initial seconds, however some differences arise at the final seconds of the accelerograms.

Plastic hinge modelling was carried out assuming the longitudinal reinforcement reduction at the pier base and therefore the corresponding decrease in the yielding moment, but keeping the plastic hinges near the foundation, although the referred longitudinal reinforcement curtailment takes place in region 2 (see Fig. 4.61).

As can be seen in Fig. 4.63, the responses for both damage model and PNL model are reasonably close, with respect to both frequency and maximum amplitude. In fact, for low level earthquake, both models give close results only in the first part of the responses, but in the final seconds the frequencies become different.

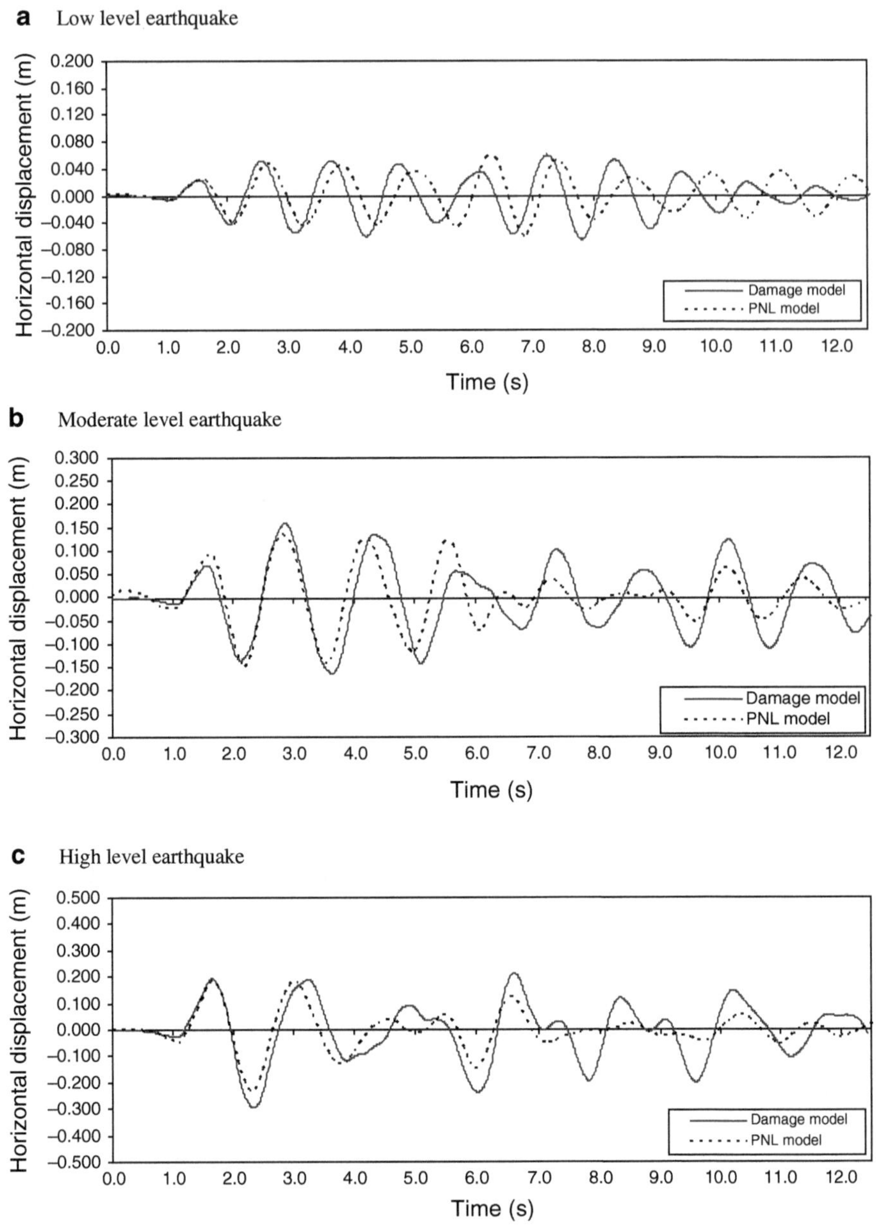

Fig. 4.63 Horizontal displacement history of the top of pier P3, comparison between damage model and plastic hinge model (with long. reinforcement reduction). (**a**) Low level earthquake; (**b**) Moderate level earthquake; (**c**) High level earthquake

For subsequent response of the numerical simulations, it was again difficult to achieve accurate results for the plastic hinge model due to the unusual reinforcement layout adopted in the design, since the longitudinal reinforcement curtailment represents an important influence on crack localization and plastic hinge formation; nevertheless, fairly good agreement was found for moderate and high level earthquake, mainly in the beginning of the accelerograms.

4.4.2.5 Concluding Remarks

The Talübergang Warth bridge was experimentally tested under asynchronous pseudo-dynamic conditions at the JRC and the results were compared against numerical simulations carried out with the damage model and a plastic hinge type model. Displacement responses obtained numerically and experimentally, for three different intensity level earthquakes, are in general close to the measured ones with respect to both frequency and maximum amplitude, indicating the capability of the damage model to provide accurate simulations of the seismic performance of the bridge although the longitudinal reinforcement curtailment represents an important influence on macro-crack localization.

The comparison between the asynchronous and synchronous response obtained with the damage model allows to conclude that maximum peak displacement is reasonable, for both moderate and high level earthquake.

For the synchronous numerical simulations with damage model and the plastic hinge model, it was quite challenging to achieve accurate results for the plastic hinge model due to the significant reduction of the longitudinal reinforcement in region 2. The results for the plastic hinge model considering the longitudinal reinforcement curtailment show good agreement with the damage model, with respect to both frequency and maximum amplitude. Nevertheless, great caution must be exercised in using the plastic hinge model regarding the aspects and singularities of the structures that are likely to affect the seismic response.

References

Antoniou S, Pinho R (2004) Development and verification of a displacement-based adaptive pushover procedure. J Earthq Eng 8(5):643–661

Applied Technology Council (ATC) (1996) Seismic evaluation and retrofit of concrete buildings, vols 1 and 2. Report No. ATC-40, Redwood City, CA

Applied Technology Council (ATC) (2005) Improvement of nonlinear static seismic analysis procedures. FEMA 440 report, Redwood City, CA

ASCE/SEI (2007) Seismic rehabilitation of existing buildings – ASCE standard 41-07. American Society of Civil Engineers, Reston

Aydinoğlu MN (2004) An improved pushover procedure for engineering practice: incremental response spectrum analysis IRSA. In: Proceedings of the international workshop: performance-based seismic design concepts and implementation. Bled, Slovenia, pp 345–356

Aydinoğlu MN, Önem G (2007) Nonlinear performance assessment of bridges with incremental response spectrum analysis (IRSA). In: ECCOMAS thematic conference on computational methods in structural dynamics and earthquake engineering. Rethymno, Greece

Bommer JJ, Martinez-Pereira A (1999) The effective duration of earthquake ground motion. J Earthq Eng 3(2):127–172

Caltrans-California Department of Transportation (2006) Seismic design criteria version 1.4. CALTRANS, Sacramento

Cardone D, Dolce M, Rivelli M (2008) Evaluation of reduction factors for high-damping design response spectra. Bull Earthq Eng 9(1):273–291

Cardone D, Perrone G, Sofia S (2011) A performance-based adaptive methodology for the evaluation of vulnerability and seismic risk of highway bridges. Bull Earthq Eng 9:1463–1498

Casarotti C, Pinho R (2006) Seismic response of continuous span bridges through fibre-based finite element analysis. J Earthq Eng Eng Vib 5(1):119–131

Casarotti C, Pinho R (2007) An adaptive capacity spectrum method for assessment of bridges subjected to earthquake action. Bull Earthq Eng 5(3):377–390

CEN (Comité Européen de Normalisation) (1994) Eurocode 8: Design provisions for earthquake resistance of structures – Part 2: Bridges (prENV 1998-2:1994). CEN, Brussels

CEN (Comité Européen de Normalisation) (2004) Eurocode 8: Design of structures for earthquake resistance – Part 1: General rules, seismic actions and rules for buildings (EN 1998-1: 2004). CEN, Brussels

CEN (Comité Europeen de Normalization) (2005) Eurocode 8: Design of structures for earthquake resistance – Part 2: Bridges, EN 1998-2:2005. CEN, Brussels

Chopra AK, Goel RK (2002) A modal pushover analysis procedure for estimating seismic demands for buildings. Earthq Eng Struct Dyn 31:561–582

Chopra AK, Goel RK, Chintanpakdee C (2004) Evaluation of a modified MPA procedure assuming higher modes as elastic to estimate seismic demands. Earthq Spectra 20(3):757–778

Computers and Structures Inc (2003) SAP2000: three dimensional static and dynamic finite element analysis and design of structures. CSI, Berkeley

Computer and Structures, Inc. (2005) SAP2000 – advanced: static and dynamic analysis finite element analysis of structures. CSI, Berkeley

Delgado P, Costa A, Delgado R (2002) A simple methodology for seismic safety assessment of bridges. In: Proceedings of the 12th ECEE, London

Delgado P, Costa A, Pinho R, Delgado R (2004) Different strategies for seismic assessment of bridges – comparative studies. In: Proceedings of the 13th world conference on earthquake engineering – 13WCEE, Vancouver, Canada

Dolce M, Cardone D, Perrone G (2007) Seismic risk assessment of highway bridges. In: Proceedings of 1st US-Italy seismic bridge workshop, 18–20 April 2007, Pavia, Italy

Fajfar P, Fischinger M (1987) Non-linear seismic analysis of RC buildings: implication of a case study. Eur Earthq Eng 1:31–43

Fajfar P, Gašperšič P, Drobnič D (1997) A simplified nonlinear method for seismic damage analysis of structures. In: Seismic design methodologies for the next generation of codes: proceedings of the international workshop Bled, Slovenia, pp 183–194

Faria R, Olive J, Cervera M (1998) A strain based plastic viscous damage model for massive concrete structures. Int J Solids Struct 35(14):1533–1558

Faria R, Vila Pouca N, Delgado R (2001) Numerical models to predict the nonlinear behaviour of bridge piers under severe earthquake loading. Deliverable no. 6, Task 3, Report No. 3/2 (VAB Project), FEUP, Porto

Giuffrè A, and Pinto P (1970) "Il comportamento del cemento armato per sollecitazione ciclice di forte intensitá." Giornale del genio civile, May (in Italian)

Isakovic T, Fischinger M (2006) Higher modes in simplified inelastic seismic analysis of single column bent viaducts. Earthq Eng Struct Dyn 35(10):95–114

Isakovic T, Fischinger M (2007) Pushover analysis of the two-span reinforced concrete bridge system. Research report, Institute of Structural Engineering, Earthquake Engineering and Construction IT, University of Ljubljana, Faculty of Civil and Geodetic Engineering, Ljubljana

Isaković T, Fischinger M (2008) Pushover analysis of a two-span multicolumn bent RC bridge, experimentally tested on three shake tables. In: Proceedings of 14th world conference on earthquake engineering. Beijing, China

Isaković T, Fischinger M (2011) Applicability of pushover methods to the seismic analyses of an RC bridge, experimentally tested on three shake tables. J Earthq Eng 15(2):303–320

Isaković T, Fischinger M, Kante P (2003) Bridges: when is single mode seismic analysis adequate? Struct Build 156(SB2):165–173

Isaković T, Pompeyo Lazaro MN, Fischinger M (2008) Applicability of pushover methods for the seismic analysis of single-column bent viaducts. Earthq Eng Struct Dyn 37(8):1185–1202. doi:10.1002/eqe.813

Jacobsen LS (1930) Steady forced vibrations as influenced by damping. Trans ASME 51:227–253

Johnson NS, Saiidi MS, Sanders DH (2006) Large-scale experimental and analytical seismic studies of a two span concrete bridge system. Report No. CCEER-06-02, Center for Civil Engineering Earthquake Research, UNR, Department of Civil and Environmental Engineering, Reno

Johnson N, Ranf RT, Saiidi MS, Sanders D, Eberhard M (2008) Seismic testing of a two-span reinforced concrete bridge. J Bridg Eng 13:173–182. doi:10.1061/(ASCE)1084-0702(2008)13:2(173)

Kappos AJ (1991) Analytical prediction of the collapse earthquake for RC buildings: suggested methodology. Earthq Eng Struct Dyn 20(2):167–176

Kappos AJ (2002) RCCOLA-90: a microcomputer program for the analysis of the inelastic response of reinforced concrete sections. Department of Civil Engineering, Aristotle University of Thessaloniki, Thessaloniki

Kappos AJ, Paraskeva TS (2008) Nonlinear static analysis of bridges accounting for higher mode effects. In: Workshop on nonlinear static methods for design/assessment of 3D structures, Lisbon, Portugal

Kowalsky MJ, Priestley MJN, MacRae GA (1995) Displacement-based design of RC bridge columns in seismic regions. Earthq Eng Struct Dyn 24(12):1623–1643

Lupoi A, Franchin P, Pinto PE (2007) Further probing of the suitability of push-over analysis for the seismic assessment of bridge structures In: Proceedings of the first US-Italy workshop on seismic design of bridges, Pavia, Italy

Maroney BA, Romstad KM, Kutter B (1993) Experimental testing of laterally loaded large scale bridge abutments. Structural engineering in natural hazards mitigation. In: Proceedings of papers presented at the structures congress '93, vol 2, ASCE, New York, pp 1065–1070

Mazzoni S, McKenna F, Scott MH, Fenves GL et al (2006) Open system for earthquake engineering simulation user command-language manual, version 1.7.3., PEER, Berkeley

Ministry of Public Works of Greece (2003) Greek seismic code – EAK 2000 (amended June 2003). Athens (in Greek)

Moschonas IF, Kappos AJ, Panetsos P, Papadopoulos V, Makarios T, Thanopoulos P (2009) Seismic fragility curves for Greek bridges: methodology and case studies. Bull Earthq Eng 7(2):439–468

Paraskeva T, Kappos A, Sextos A (2006) Development and evaluation of a modal pushover analysis procedure for seismic assessment of bridges. Earthq Eng Struct Dyn 35(10):1269–1293

Pinho R, Casarotti C, Antoniou S (2007) A comparison of single-run pushover analysis techniques for seismic assessment of bridges. Earthq Eng Struct Dyn 36(10):1347–1362

Pinto A, Pegon P, Magonette G, Molina J, Buchet P, Tsionis G (2002) Pseudodynamic tests on a large-scale model of an existing RC bridge using non-linear substructuring and asynchronous motion. EUR 20525 EN, EC, Joint Research Centre, ISIS

Priestley MJN, Seible F, Calvi GM (1996) Seismic design and retrofit of bridges. Wiley, New York, Chichester, Brisbane, Toronto, Singapore

SAC Joint Venture (1997) Develop suites of time histories. Project task: 5.4.1, Draft Report, March 21, Sacramento

SeismoSoft (2006) SeismoStruct – a computer program for static and dynamic nonlinear analysis of framed structures available online from http://www.seismosoft.com

Sextos A, Pitilakis K, Kappos AJ (2003) Inelastic dynamic analysis of RC bridges accounting for spatial variability of ground motion, site effects and soil-structure interaction phenomena. Part 1: Methodology and analytical tools. Earthq Eng Struct Dyn 32(4):607–627

Takahashi Y (2009) http://opensees.dpri.kyoto-u.ac.jp/japanese/binaries.html (Accessed September 2009)

UNR (2008) http://nees.unr.edu/4-spanbridges/index.htm (Accessed September 2008)

Vila Pouca N (2001) Simulação Numérica da Resposta Sísmica de Elementos Laminares em Betão Armado. Dissertation, FEUP (http://ncrep.fe.up.pt/web/artigos/NVilaPouca_PhD_Thesis.pdf) (in Portuguese)

Vila Pouca N, Faria R, Delgado R (2002) Numerical simulation of the seismic behaviour of Talübergang Warth bridge. In: Proceedings of the 12th ECEE, London

Chapter 5
Conclusions and Recommendations

Andreas J. Kappos and Tatjana Isaković

5.1 Modelling Issues

From the wealth of information presented in Chap. 2, it is clear that the bridge engineer has at his/her disposal a set of powerful analysis tools that can be used for the design or assessment of any bridge type. The potential of these tools, when properly utilised, was revealed by their success in predicting the response of bridges tested under high levels of earthquake actions, that caused substantial amounts of inelasticity, as described in Sect. 4.4 of the book. The material presented in Chap. 2 offers to researchers and designers all the necessary information regarding the available models for the various parts of the bridge (deck, bearings and shear keys, isolation and energy dissipation devices, piers, foundation members), as well as tools for modelling the dynamic interaction between piers, foundation and soil, and the abutment-embankment-superstructure system. It also provided information on important parameters that help ensuring that inelastic analysis of bridge earthquake response is conducted properly. It has to be emphasised in this respect that the power and versatility of the analysis tools also makes the results particularly sensitive to improper application.

It is not a coincidence that the largest part of Chap. 2 was devoted to modelling of piers, as these members are both the ones wherein energy dissipation through plastic hinging is intended to occur (unless a seismic isolation system is used), and those whose inelastic response is relatively easier to model in inelastic analysis

A.J. Kappos (✉)
Department of Civil Engineering, Aristotle University of Thessaloniki,
54124 Thessaloniki, Greece
e-mail: ajkap@civil.auth.gr

T. Isaković
Faculty of Civil and Geodetic Engineering, University of Ljubljana, Jamova 2,
1000 Ljubljana, Slovenia
e-mail: tisak@ikpir.fgg.uni-lj.si

A.J. Kappos et al. (eds.), *Seismic Design and Assessment of Bridges: Inelastic Methods of Analysis and Case Studies*, Geotechnical, Geological and Earthquake Engineering 21,
DOI 10.1007/978-94-007-3943-7_5, © Springer Science+Business Media Dordrecht 2012

(compared, for instance, to the abutment-backfill system or some foundation types or, indeed, some types of joints). Having said this, it is also clear from the material presented in this chapter, that proper modelling of the other components of the bridge, even those that are typically assumed to remain elastic during the seismic excitation (such as prestressed concrete decks) is also important, since, through their stiffness characteristics, affect the dynamic characteristics of the bridge and the way seismic actions are transferred to the yielding members. Of great importance is also the modelling of the various connections in the bridge system, i.e. those between piers and deck, abutments and deck and, in the common case (especially in the transverse direction) that the movement of the deck is restrained at the location of the abutment, the proper modelling of the response of the abutment-backfill system. Various aspects of the latter critical modelling issue were presented in Sect. 2.7 and approaches of varying complexity were presented in detail. It is noted here that in a practical context and when the main objective of the analysis is the response of the bridge itself (rather than that of the surrounding ground), the recommended solution is to carry out an independent analysis of the abutment-embankment system, determine its resistance curves (in all relevant directions), and use them to describe the nonlinear response of the equivalent springs to which the bridge model will be connected (e.g. see Table 2.10). If such an analysis cannot be afforded, the properties of these springs can be defined on the basis of simplified guidelines from the literature. For pier-foundation-soil interaction the existing literature is more mature and often it is not necessary to carry out separate analysis of the system to derive the nonlinear properties of the soil-foundation dynamic impedance to be introduced in the bridge model, especially when surface foundations are used; in these cases information from the literature can be used to account approximately for the interaction with the surrounding soil. Furthermore, the designer has to be able to distinguish between the static and dynamic stiffness of the soil-foundation-superstructure system, because this is eventually related to the different properties of the spring and dashpot constants to be used for different types of analysis (i.e. nonlinear response history or nonlinear static).

Arguably, the largest room for choosing between alternative approaches, none of which is a clear preferred candidate, is the case of piers. In addition to presenting the various available models, Chap. 2 also presented several examples of applying different models to the same piers and comparing the calculated response resulting from each model. It is clear from these examples that when simplicity and computational cost (mainly in the sense of preparation, testing, and verification of the model, i.e. in terms of human, rather than computer, resources) are the main consideration, the 'classic' one-component model (lumped plasticity or point hinge model) is the preferred choice. However, there are instances that more detailed information on pier damage is required, in which case other, more refined models should preferably be selected among those presented in Sect. 2.5; the condensed critical comments on each model included in Table 2.2 would be useful in selecting an appropriate model.

5.2 Reliability of Inelastic Analysis Methods

All currently available inelastic analysis methods were presented in Chap. 3 in a depth sufficient for understanding and applying them, using of course, the appropriate software. From the application of the methods presented in Chap. 3 to a fairly large number of case-studies, involving bridges with different length and configuration, the limitations and the relative efficiency of the individual methods were brought into light. The studies of bridges designed according to the modern seismic codes, presented in Sects. 4.3.1, 4.3.2 and 4.3.3, confirmed some trends previously identified in the literature regarding the applicability of the standard single mode methods. These methods, (a typical example of which is the N2 method, can be used when the higher modes do not considerably influence the response of a bridge and when the fundamental mode shape does not considerably change based on the seismic intensity. Actually, these are the basic assumptions of the majority of the single mode methods. While they are almost always valid in the longitudinal direction of typical bridges, in the transverse direction there are many cases where these assumptions are not fulfilled or they are only satisfied up to a certain level of seismic intensity (and associated degree of inelasticity).

In general, the single mode methods are suitable for the analysis of bridges where the effective mass of the fundamental mode of the structure in the direction examined, exceeds approximately 80% of the total mass and the influence of higher modes is small. In this way the first assumption of the method related to the higher modes can be tested. The second assumption related to the invariability of the predominant mode can be tested based on an appropriate index, such as that proposed in Isaković et al. (2003).

It can be concluded that single mode methods (such as the N2 method) can be primarily used for short-to-medium length bridges, which are supported by relatively long columns, and where the displacements of the superstructure above the abutments are restrained (i.e., superstructure can be assumed as pinned in the transverse direction). Subject to a number of conditions, these methods can be also used for the analysis of the same type of bridges, supported by short and quite stiff columns; however, only in case that the latter are not located in the central part of the bridge (an inappropriate design practice, which, nevertheless can be found in some existing older bridges). In such bridges the accuracy of the N2 method in general increases proportionally to the seismic intensity.

The results of the analysis of short and medium length bridges, which are pinned above the abutments, and supported by short and stiff columns located away from their centre, are accurate regardless of the seismic intensity when one of the multimodal pushover methods are used, i.e. modal pushover (MPA), modified MPA, inelastic response spectrum (IRSA), or adaptive capacity spectrum (ACSM). The majority of the multimode methods can be also used for the analysis of all types of short bridges pinned at the abutments and for the analysis of long bridges, which are not supported on very short and stiff columns in the central part of the bridge. The accuracy of these methods, however, can be different and can

Table 5.1 Recommended types of inelastic analysis

Type of bridge	Single-mode methods	Multi-mode methods: Non-adaptive	Multi-mode methods: Adaptive	Nonlinear response history analysis
Response is governed predominantly by one mode, which does not considerably change:	X			
Short bridges on moderate to stiff soil, pinned at the abutments, and not supported by very short columns				
The influence of higher modes is limited and their shape does not considerably change when the seismic intensity is increased:	X	X		
Short bridges pinned at the abutments, supported by short side and long central columns				
Considerable influence of higher modes, that do not significantly changed their shape:		X	X	
Long bridges (or curved) without very short central columns				
Considerable influence of one or a few number of modes, which significantly change the shape:			X	
Short bridges with roller supports at the abutments				
Considerable influence of higher modes, which significantly change their shape when the seismic intensity is changed:				X
Short or long bridges supported by very short central and higher side columns				

depend on the seismic intensity, particularly when the method can be characterized as non-adaptive (a typical representative being the MPA method) since the latter group of methods cannot take into account modification of the mode shapes due to seismic loading. Such changes of the mode shapes can be expected particularly in cases where torsional stiffness of the bridge becomes lower than the translational one. Typically, this can be observed in short bridges with relatively small number of columns and roller supports above the abutments (see Sect. 4.4). In these types of bridges the influence of the higher modes is usually limited, but the shape of the predominant mode can considerably change when the seismic intensity is increased; for these bridges the use of adaptive methods (such as the IRSA) is recommended.

Finally, multimode pushover methods of any type should be applied with caution in the analysis of bridges independently of their length, especially in case that they are centrally supported by short and stiff columns. The reason is that, the response of these types of bridges is very complex since it is significantly influenced by higher modes, which substantially change for increasing seismic intensity. Because of this very complex response, non-linear response history analysis is recommended in this case.

The observations presented above are summarized in Table 5.1. For each group of bridges the recommended types of analysis are marked with an X.

As made clear in the Introduction, it was *not* the aim of this book to suggest the 'best' method of seismic assessment of bridges, for the simple reason that none of the existing methods would qualify as such. As confirmed by the numerous case-studies of Chap. 4, selection of analysis method is dependent on the configuration of the bridge studied and is also influenced by the available software. Indeed, one should note that practical application of any of the advanced pushover techniques described in this book is feasible only if it is supported by the appropriate software, otherwise it remains within the realm of research.

Reference

Isaković T, Fischinger M, Kante P (2003) Bridges: when is single mode seismic analysis adequate? Struct Build 156(SB2):165–173

Index

A.J. Kappos et al. (eds.), *Seismic Design and Assessment of Bridges: Inelastic Methods of Analysis and Case Studies*, Geotechnical, Geological and Earthquake Engineering 21, DOI 10.1007/978-94-007-3943-7, © Springer Science+Business Media Dordrecht 2012

MIX
Papier aus verantwortungsvollen Quellen
Paper from responsible sources
FSC® C105338

If you have any concerns about our products,
you can contact us on
ProductSafety@springernature.com

In case Publisher is established outside the EU,
the EU authorized representative is:
Springer Nature Customer Service Center GmbH
Europaplatz 3, 69115 Heidelberg, Germany

Printed by Libri Plureos GmbH
in Hamburg, Germany